IMAGE EVALUATION
TEST TARGET (MT-3)

6"

Photographic
Sciences
Corporation

23 WEST MAIN STREET
WEBSTER, N.Y. 14580
(716) 872-4503

**CIHM
Microfiche
Series
(Monographs)**

**ICMH
Collection de
microfiches
(monographies)**

Canadian Institute for Historical Microreproductions / Institut canadien de microreproductions historiques

© 1993

Technical and Bibliographic Notes / Notes techniques et bibliographiques

The Institute has attempted to obtain the best original copy available for filming. Features of this copy which may be bibliographically unique, which may alter any of the images in the reproduction, or which may significantly change the usual method of filming, are checked below.

L'Institut a microfilmé le meilleur exemplaire qu'il lui a été possible de se procurer. Les détails de cet exemplaire qui sont peut-être uniques du point de vue bibliographique, qui peuvent modifier une image reproduite, ou qui peuvent exiger une modification dans la méthode normale de filmage sont indiqués ci-dessous.

☐ Coloured covers/
Couverture de couleur

☐ Covers damaged/
Couverture endommagée

☐ Covers restored and/or laminated/
Couverture restaurée et/ou pelliculée

☐ Cover title missing/
Le titre de couverture manque

☐ Coloured maps/
Cartes géographiques en couleur

☐ Coloured ink (i.e. other than blue or black)/
Encre de couleur (i.e. autre que bleue ou noire)

☐ Coloured plates and/or illustrations/
Planches et/ou illustrations en couleur

☐ Bound with other material/
Relié avec d'autres documents

☐ Tight binding may cause shadows or distortion along interior margin/
La reliure serrée peut causer de l'ombre ou de la distorsion le long de la marge intérieure

☐ Blank leaves added during restoration may appear within the text. Whenever possible, these have been omitted from filming/
Il se peut que certaines pages blanches ajoutées lors d'une restauration apparaissent dans le texte, mais, lorsque cela était possible, ces pages n'ont pas été filmées.

☐ Additional comments:/
Commentaires supplémentaires:

☐ Coloured pages/
Pages de couleur

☐ Pages damaged/
Pages endommagées

☐ Pages restored and/or laminated/
Pages restaurées et/ou pelliculées

☑ Pages discoloured, stained or foxed/
Pages décolorées, tachetées ou piquées

☑ Pages detached/
Pages détachées

☑ Showthrough/
Transparence

☑ Quality of print varies/
Qualité inégale de l'impression

☐ Continuous pagination/
Pagination continue

☐ Includes index(es)/
Comprend un (des) index

Title on header taken from:/
Le titre de l'en-tête provient:

☐ Title page of issue/
Page de titre de la livraison

☐ Caption of issue/
Titre de départ de la livraison

☐ Masthead/
Générique (périodiques) de la livraison

This item is filmed at the reduction ratio checked below/
Ce document est filmé au taux de réduction indiqué ci-dessous.

10X		14X		18X		22X		26X		30X	
	12X		16X		20X		24X		28X		32X

The copy filmed here has been reproduced thanks to the generosity of:

National Library of Canada

The images appearing here are the best quality possible considering the condition and legibility of the original copy and in keeping with the filming contract specifications.

Original copies in printed paper covers are filmed beginning with the front cover and ending on the last page with a printed or illustrated impression, or the back cover when appropriate. All other original copies are filmed beginning on the first page with a printed or illustrated impression, and ending on the last page with a printed or illustrated impression.

The last recorded frame on each microfiche shall contain the symbol → (meaning "CONTINUED"), or the symbol ▽ (meaning "END"), whichever applies.

Maps, plates, charts, etc., may be filmed at different reduction ratios. Those too large to be entirely included in one exposure are filmed beginning in the upper left hand corner, left to right and top to bottom, as many frames as required. The following diagrams illustrate the method:

L'exemplaire filmé fut reproduit grâce à la générosité de:

Bibliothèque nationale du Canada

Les images suivantes ont été reproduites avec le plus grand soin, compte tenu de la condition et de la netteté de l'exemplaire filmé, et en conformité avec les conditions du contrat de filmage.

Les exemplaires originaux dont la couverture en papier est imprimée sont filmés en commençant par le premier plat et en terminant soit par la dernière page qui comporte une empreinte d'impression ou d'illustration, soit par le second plat, selon le cas. Tous les autres exemplaires originaux sont filmés en commençant par la première page qui comporte une empreinte d'impression ou d'illustration et en terminant par la dernière page qui comporte une telle empreinte.

Un des symboles suivants apparaîtra sur la dernière image de chaque microfiche, selon le cas: le symbole → signifie "A SUIVRE", le symbole ▽ signifie "FIN".

Les cartes, planches, tableaux, etc., peuvent être filmés à des taux de réduction différents. Lorsque le document est trop grand pour être reproduit en un seul cliché, il est filmé à partir de l'angle supérieur gauche, de gauche à droite, et de haut en bas, en prenant le nombre d'images nécessaire. Les diagrammes suivants illustrent la méthode.

1	2	3

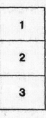

1	2	3
4	5	6

OUTLINES

OF

CHRONOLOGY;

FOR THE USE OF SCHOOLS.

EDITED BY

MRS. GORDON.

Montreal:

PRINTED AND PUBLISHED BY JOHN LOVELL,

ST. NICHOLAS STREET;

Toronto:

W. C. F. CAVERHILL, BOOKSELLER AND STATIONER,

87 YONGE STREET.

1859.

THE follow
printed for
a convenie
intended n
Editor wil
beyond the
A similar w
self, in the
It may pos
perhaps, m
ing portion
laudable p
knowledge
and acquire
In the ea
Tables com
ployed. Ar
history wou
memory, an
a degree of
in fact. Bu
thing like a
to secure it

PREFACE.

THE following elementary outline of chronological science is printed for the purpose of furnishing the Editor's pupils with a convenient help to their historical studies. But, although intended mainly to supply a felt want in her own seminary, the Editor will be gratified if it shall also be found acceptable beyond the limits for which it is more immediately destined. A similar want may have been felt by others, engaged, like herself, in the labours of tuition; and the present attempt to supply it may possibly receive from them a friendly welcome. Nor, perhaps, may it prove altogether unacceptable to that interesting portion of the youthful community who have formed the laudable purpose of preserving and extending at home the knowledge of which at school they have laid the foundations and acquired the love.

In the early, and essentially conjectural, dates with which the Tables commence, it is only round numbers that have been employed. Any departure from these at that period of the world's history would both be an unnecessary tax upon the pupil's memory, and have a positive tendency to mislead, by indicating a degree of exactness in our knowledge which has no existence in fact. But after we come down to a period at which something like accuracy is really attainable, pains have been taken to secure it; and it is believed that no very glaring errors will

be found to have been committed. The events selected for
notice are either such as are in themselves of prominent historical importance, or such as, from their connexion with others
of that description, seem calculated to serve as convenient aids
to the memory.

It is almost superfluous to observe, that in an introductory
manual of this sort, where utility with reference to an educa-
tional object is the sole end in view, the Editor could have no
hesitation in copiously availing herself of the labours of others.
The sources from which she has compiled will indeed be suffi-
ciently obvious to all who are really acquainted with modern
works upon the subject. But it may do no harm to mention
that nothing whatever has been borrowed from any work in the
English language prepared for the use of schools; as, indeed,
had the Editor been acquainted with any manual suitable for
her purpose, she would have greatly preferred using what lay
ready to her hand, to bestowing on the preparation of the following pages an amount of time and labour utterly disproportioned
to the slenderness of their bulk and pretensions. She will, however, regret neither her hours nor her trouble, if the little
manual shall be found but half as useful as she hopes it is calculated to become.

ARGYLE TERRACE,
    Montreal, July, 1859.

INTRODUCTION.

§ 1.

CHRONOLOGY (from the Greek words *chronos*, time, and *logos*, a discourse) is the science which teaches the measurement and division of Time; and, as a result of this knowledge, enables us to arrange the events of History in the order of their sequence, and to ascertain the intervals of time between them.

To give a complete definition of the notion of *time*, has been found one of the puzzles of metaphysics. "Si non rogas, intelligo"* still remains as good an answer, to the question of what time is, as we are ever likely to obtain. Without troubling ourselves, therefore, about a definition, we may observe, that, as long as we are awake, there is a train of ideas which constantly *succeed* one another in the mind;—that, when

---

* Mr. Locke translates this, "The more I set myself to think of it, the less I understand it"; but the words might also be rendered, "I understand what it is, but I can't answer your question." We really have a perfect conception of what time *is*, as well as space and motion, although we are unable to define any of the three. Our inability to arrive at a definition arises, not from their obscurity, but from their extreme clearness. It was an observation of Descartes, that it is vain, and even dangerous, to attempt a definition of evident things, because in such cases we are apt to mistake a definition of the *word* for a definition of the *thing*.

we reflect upon these appearances, we are furnished with the idea of *succession ;*—and that the distance between any two parts of that succession, or between the appearance of any two ideas in our minds, is what we call *duration.* Thus, the succession of ideas presents a consciousness from which we derive the perception of a certain lapse of time.

§ 2.

That an actual succession of ideas is requisite to our perception of time, is manifested by the circumstance, that, when that succession ceases, our perception of duration ceases with it. Thus, in profound sleep without dreaming, we have no perception whatever of duration. Having gone to sleep at night, we know, it is true, on waking in the morning, that a certain definite interval has elapsed; but we derive this knowledge by inference from external phenomena, and not from consciousness. We see that the darkness of night has changed to the light of day,—that the sun, which was below the horizon, is above it; and we know by experience that these changes are produced only *in a certain interval of time ;* whence we conclude that such an interval must have elapsed since we fell asleep. But if we fall asleep in the evening, and do not awaken until the next day but one, we are unconscious of the lapse of more than one night.

§ 3.

Although a succession of ideas floating at hazard through the mind, or excited casually and without regularity by external objects, produces a *perception* of time, it does not afford a *measure* of it. A series of agreeable thoughts has the effect of making time appear to pass with rapidity, whilst a series of painful or disagreeable impressions has the effect of making an equal portion of time seem very much longer. The series of events or perceptions which would supply a measure of time must be absolutely uniform and regular. In such case, the number of repetitions of the same event or phenomenon found between any two points of the series becomes the measure of the interval of time which has elapsed between them.

§ 4.

The series of phenomena which have been most universally adopted by mankind as measures of time, are those regularly recurring phenomena, the (real or apparent) motions of the heavenly bodies, especially those of the sun (earth) and moon, which are easily observable by all, and never cease to be reproduced with unvarying uniformity. Thus the revolutions of the earth upon its axis, of the moon round the earth, and of the earth round the sun, give rise to those natural divisions of time with which we are familiar under the denominations of *days*, *months*, and *years*. The other divisions of time are of a more artificial and arbitrary nature.

§ 5.

The term *day* has two distinct significations: it may mean either the *natural* or the *civil* day. The former, as opposed to night, means the interval during which

the sun remains above the horizon; and at all parts of
the earth, except the equator, it varies in length according to the season of the year: the latter embraces
the period of a single complete revolution of the earth
upon its axis, and consequently includes an entire day
and night (Gr. *nuchthemeron*, Lat. *noctiduum*).

§ 6.

The day, whether natural or civil, is divided into
hours; and in all nations, from time immemorial, this
division of the day has been a duodecimal division.
Some peoples have counted the hours consecutively
from one to twenty-four; others have divided the day
into two series of twelve hours each. It is probable
that amongst the nations of antiquity the customary
division of the natural day from sunrise to sunset, and
of the natural night from sunset to sunrise, was each
into twelve equal parts, called respectively hours of
the day and hours of the night. It is evident that the
diurnal hours could have been equal in length to the
nocturnal hours only at the equinoxes; that from the
vernal to the autumnal equinox the diurnal were longer
than the nocturnal hours, and from the autumnal to
the vernal equinox the nocturnal were longer than the
diurnal hours. The hours, both diurnal and nocturnal,
were also subject to continual variations of length.
From the first day of winter, or the shortest day, to the
first day of summer, or the longest day, the diurnal
hours constantly increased, and the nocturnal hours
constantly decreased in length; and from the first day
of summer to the first day of winter, the nocturnal

hours constantly increased in length, whilst the diurnal hours constantly diminished. Inconvenient as such a mode of measuring the time must have been, according to our present notions, it is probable that it continued to be employed in Europe during the greater portion of the Middle Ages; and that the division into hours of uniformly equal length did not take place, except for astronomical purposes (equinoctial hours), until the invention of mechanical clocks towards the twelfth century of the Christian era.

§ 7.

The moment at which the day commences is a point in regard to which there has been much disagreement in the practice of different nations. The Babylonians or Chaldeans began their day at sunrise, and divided it and the night into twelve hours each. They also invented sun-dials and water-clocks of a rude description. The Hebrews fixed the commencement of their day at sunset: their ancient division of it was into evening and morning, noon and night—the night being also divided into watches: it was from the Babylonians that they first learned the division into hours. The ancient Greeks also commenced their day at sunset. Before the time of Herodotus (B. C. 450), they appear to have learned from the Babylonians both the division of the day into hours, and the use of the sun-dial; but neither of these were in common use amongst them until long afterwards. To denote the time of day, they employed expressions borrowed from the common functions of life; and, for ascertaining the four watches into which they divided the night, they made use of a rude water-clock called a clepsydra. The Romans divided the natural day and the natural night into twelve hours each; but as their civil day began at midnight—the seventh nocturnal hour—their nocturnal hours were thus distributed over two civil days. Like the Greeks, they divided the night into four watches (*vigiliae*), in the regulation of which they were guided by observing the heavenly bodies, or by the clepsydra. The position of the sun served to indicate to them the time of day. About three centuries before Christ they were acquainted with the sun-dial: the water-clock was introduced among them by Scipio Nasica in the year B. C. 164. The Italians, even to the present time, divide the day into twenty-four successive hours, reckoned continuously from sunset to sunset. Thus, at an hour before sunset, it is twenty-three o'clock; at two hours before sunset, it is twenty-two o'clock, and so on. According to this system, the hour of sunrise varies from day to day, and from season to season; but the hour of sunset is constant, being twenty-four o'clock. It is evident that a clock to indicate such time must be set from day to day, or at least from week to week, since each succeeding sunset would be constantly later than the previous one during one-half the year, and constantly earlier during the other half. The English, French, Germans, and the moderns generally, commence the day at midnight and divide it into two equal series of twelve hours, so that midday and midnight are equally twelve o'clock. According to this system of reckoning, it is necessary, whenever an hour

ade use of a
he Romans
night into
y began at
r nocturnal
lays. Like
ur watches
rere guided
clepsydra.
to them the
Christ they
rater-clock
ica in the
esent time,
ive hours,
:. Thus, at
o'clock; at
lock, and
of sunrise
to season;
venty-four
icate such
from week
ld be con-
ie-half the
half. The
generally,
t into two
and mid-
ng to this
r an hour

is named, to indicate its relation to noon. The hours before noon are indicated by the letters A. M., and those after noon by P. M., being the initials of the Latin words *ante meridiem* (before midday), and *post meridiem* (after midday). Modern astronomers divide the day into twenty-four successive hours reckoned from noon to noon. Thus, according to their manner of reckoning, twenty minutes and a half after ten o'clock in the morning would be 22h. 20m. 30s. Civil or common time, therefore, is half a day before astronomical time. Thus, for example, the first day of the year 1859, according to civil reckoning, commenced at the moment of midnight between the 31st December, 1858, and 1st January, 1859; but, according to astronomical reckoning, it commenced at midday on 1st January, 1859. It follows, therefore, that the twelve hours which preceded the noon of 1st January, 1859, were, according to astronomical reckoning, the last twelve hours of the year 1858.

§ 8.

Next to the alternations of day and night, the changes of the moon form the most striking appearance amongst the heavenly bodies; and as the regularity of their occurrence must soon have attracted attention, we are not surprised to find that in the earliest ages they were used as a measure of time.

The sun (apparently) and the moon (really) move round the celestial sphere in the same direction; but the moon moves more than thirteen times as fast as the sun, and consequently makes more than thirteen revolutions of the heavens whilst the sun makes one. The moon is, therefore, constantly either departing from or approaching to and overtaking the sun. At the moment it overtakes the sun, it is said to be in *conjunction*, and is called *new moon*. At the moment it is in the opposite part of the heavens, it is in *opposition;* and as it then presents its enlightened hemisphere directly towards the earth, it appears with a complete circular disc, and is called *full moon*. When it is midway betwixt conjunction and opposition, it is said to be in its quarters : it then appears as an enlightened semicircle, and is called *half moon*. The time which it takes to make one complete revolution of the heavens is called its *period*, and is found by exact observation to amount to 27 days 7 hours 43 minutes and $11\frac{4}{10}$ seconds.

The interval between two successive conjunctions of the moon with the sun, or between two successive new moons, is greater than the moon's period. If we suppose the sun and moon to start together from conjunction, the moon, moving more than thirteen times as fast, immediately gets before the sun, and returns to the point from which she had started in 27d. 7h. 43m. and $11\frac{4}{10}$s., as we have just seen : but during this time the sun also has been (apparently) moving in the same direction, and the next conjunction, or new moon, cannot therefore take place until the moon overtakes him, which it will still require somewhat more than two days to accomplish. By the most exact observations and calculations it has been found that the interval between two successive conjunctions is 29d. 12h. 44m. $2\frac{89}{1000}$s. This interval is called a *lunation*.

ı makes one.
er departing
the sun. At
said to be in
he moment it
is in *opposi-*
l hemisphere
ı a complete
en it is mid-
It is said to
enlightened
me which it
the heavens
observation
ı and $11\frac{4}{10}$

nnctions of
essive new
If we sup-
n conjunc-
nes as fast,
rns to the
43m. and
s time the
the same
ıoon, can-
akes him,
two days
s and cal-
between
1. $2\frac{88}{1000}$s.

§ 9.

The moon's *period* is not well suited for a measure of civil time, because the moment which terminates one period and begins the next is not marked by any conspicuous and generally observable phenomenon, and can only be ascertained by astronomers. But the recurrence of the *lunation*, and even of its fractional parts, is marked by phenomena so striking and so universally observable even without instruments, that in all ages and countries we find it used by common consent as a measure of time. It received the appellation of *month* (Anglo-Saxon *monath*, from *mona*, the moon): but the civil month, being now arranged with reference to the year (see § 10), no longer corresponds to the lunation, nor does it retain any reference to the moon beyond the name.

A natural division of the lunation into four quarters corresponding to the changes of the moon, would make each of these quarters correspond pretty nearly to a period of seven days ; and such has probably enough been the origin of the *week* amongst various ancient nations, including the Chinese, the ancient Peruvians, and the ancient Germans. The week of the Christian nations, however, is derived from the Jews, amongst whom, by the setting apart of the Sabbath, it possessed from the first a religious signification. One of the chief Jewish festivals, namely, the feast of Pentecost (the thanksgiving for the harvest), was determined by a cycle of weeks, and was therefore often called the Feast of Weeks. But in the affairs of common life the Jews seem to have reckoned more usually by days than by

weeks until after their return from the captivity. Neither in the Old nor in the New Testament do we find
names for the different week days.   These names would
appear to have had originally an astrological meaning,
referring to the different planets under the influence of
which the first hour of each particular week-day was
imagined to stand, as the Sun, the Moon, Mars, Mercury,
Jupiter, Venus, and Saturn. This astrological week was,
about the commencement of the Christian era, introduced amongst the Greeks (who until then had divided
their months into *decades*, or periods of ten days), and
amongst the Romans (who previously had had a week of
eight days) ; and by the time that the Christian week,
adopted from the Jews, was officially recognised by
Constantine, the heathen names for the week-days had
taken such firm root that they continued in common
use.   Besides the name of Sunday, however, (*dies solis*,)
the Christians gave the first day of the week an appellation, *dies dominicus* (the Lord's day), which has been
retained in the modern Romanic languages (Ital. *Do-
menica ;* French, *Dimanche*).   In these languages, however, for the week-days from Monday to Friday inclusive,
the original names were retained, the seventh day re-
ceiving the name of the Sabbath.   Thus,

LATIN.	ITALIAN.	FRENCH.
Dies Lunae.	Lunedi.	Lundi.
" Martis.	Martedi.	Mardi.
" Mercurii.	Mercordi.	Mercredi.
" Jovis.	Giovedi.	Jeudi.
" Veneris.	Venerdi.	Vendredi.
" Saturnii (Sab- bati).	Sabato.	Samedi.

ptivity. Nei-
nt do we find
names would
ical meaning,
e influence of
reek-day was
ars, Mercury,
cal week was,
n era, intro-
had divided
n days), and
1ad a week of
ristian week,
1cognised by
eek-days had
in common
r, (dies solis,)
ek an appel-
1ich has been
es (Ital. Do-
guages, how-
lay inclusive,
'enth day re-

'RENCH.
1di.
:di.
:credi.
di.
1dredi.
1edi.

The English names for the days of the week have come down from the Anglo-Saxons.

Previously to the introduction of Christianity amongst them, the ancient Germans seem to have had a week of their own, and to have given names to its days under Roman influence, which may probably enough have reached them through Gaul. For Sunday and Monday the Roman names were retained: for the other days they selected the names of those German deities which seemed to bear most resemblance to the corresponding gods of Rome. To the Roman Mars corresponded the Teutonic Ziu, whence dies Martis became Ziu's-day, Tuesday; to Mercury corresponded the Teutonic Wodan (Scand. Odin), whence dies Mercurii became Wodan's-day, Wednesday; dies Jovis became the day of Thor, the god of thunder, Thor's-day, Thursday; dies Veneris became the day of Frîa, the spouse of Wodan, Frîa's-day, Friday; dies Saturnii, Saturn's-day, Saturday, was retained in many of the Teutonic languages, and amongst others in the Anglo-Saxon; but in the Scandinavian dialects it became *laugar dagr, löverdag,* or *lördag,* equivalent to *bathing* or *washing-day.*

§ 10.

Although moons or months afforded a convenient mode of measuring time amongst the earliest peoples, it could not, in the course of time fail to strike them that a more convenient mode of reckoning and dividing time, not only for the more regularly recurring labours of the field, but even for the pursuits of pasturage, hunting, and fishing, arose from the changes of the seasons. Thence would originate the division of time into *years*, or those intervals in the course of which the sun makes a complete revolution of the heavens, and the seasons are periodically reproduced.

The ascertainment of the exact period occupied by such a revolution was a matter of some difficulty. In their first attempt at the establishment of the annual standard, the Egyptians gave the year 360 days, which they divided into twelve periods of 30 days each. These periods we may, if we please, term months; but we must bear in mind that they were months which, like our modern civil months, had no reference whatever to the moon. They subsequently added five supplementary days to their year: these five days (the addition of which to the original year is attributed to one of their gods or heroes called by the Greeks Hermes Trismegistos) were intercalated betwixt the termination of the old year and the commencement of the new. Amongst other nations, the practice of reckoning by moons had taken such root before they began reckoning by years, that, when they adopted the latter mode of reckoning, they assumed the year to consist of a certain number of lunar months. The year of the Jews consisted of twelve lunar months, to which, however, in order to establish an approximative harmony with the solar year, they occasionally added a thirteenth. The new month commenced with the first appearance of the new moon in the evening twilight. If the weather was not clear enough to allow the new moon to be seen when it was looked for, the new month was reckoned to commence at the end of thirty days from

ion of time
' which the
avens, and

cupied by
culty. In
he annual
ays, which
ays each.
nths; but
hs which,
ice what-
five sup-
lays (the
ibuted to
s Hermes
termina-
the new.
oning by
reckon-
er mode
ist of a
he Jews
owever,
ny with
rteenth.
earance
If the
v moon
th was
rs from

the commencement of the old one. At the end of
twelve lunar months, the question whether a new year
were to commence or a thirteenth month were to be
added, depended on the circumstance of whether the
growing crop of barley were so far advanced towards
maturity that, by the middle of the first month, afterwards called Nisan, which fell in the time of the vernal
equinox, the sheaf of the first fruits could be offered to
Jehovah according to the command of Moses. This
fluctuating mode of measuring their years was retained
by the Jews not only throughout the period of the
Babylonian captivity, but down to the destruction of
Jerusalem. The division of the year amongst the
Greeks is involved in some obscurity; and this obscurity is increased by the circumstance that the different
Hellenic States were at variance with each other in
regard both to the number and the lengths of their
months, as well as to the period at which the year commenced. We know, however, that they early attained
a very precise knowledge of the period of the lunar
phases, and that these served as the basis of their
chronometric system. They estimated the lunation at
29¼ days (which is within three quarters of an hour of
its actual length), and this period they took for their
month. For the sake of convenience they made the
months consist of 29 and 30 days alternately; and
twelve of these months composed their year. Such a
year, however, consisting of only 354 days, deviated
from the periodical return of the seasons by more than
11 days, so that after the lapse of three years the seasons were put back more than a month, and after the

B

lapse of eighteen years they were actually reversed,—midsummer taking the place of midwinter, and *vice versa*. But the return of the seasons constituted so obvious and so natural a measure of the year, and was so intimately connected with the business of life, and especially with agriculture, that no standard which greatly varied from it could long be maintained. Attempts, therefore, were early made amongst the Greeks to bring their series of twelve months into harmony with the period of the seasons, and finally their several lengths were so adjusted as to produce a total of 365 days,—an interval so nearly corresponding to the true succession of the seasons that an age must elapse before any important discordance would be rendered manifest.

The Roman mode of measuring the year was originally very rude; and it continued very imperfect down to the time of the great reformation of the calendar effected by Julius Cæsar. The Roman year is supposed to have consisted originally of 304 days, portioned amongst ten months,* and to have been extended (in

* The first four months of the Roman year were named *Martius, Aprilis, Majus*, and *Junius*. The name Martius is derived from the god Mars, the supposed father of Romulus. Aprilis is derived by Ovid from the Latin word *aperire*, to open, in allusion to the opening vegetation of spring: more probably, however, it comes by metathesis from *Parilis*, the equivalent of Palilis, because on the 21st of that month, which was the supposed anniversary of the foundation of the city by shepherds, were celebrated the *Parilia* or *Palilia*, the festival of the shepherd god Pales. The true etymologies of Majus and Junius are uncertain, although various conjectural ones have been assigned. The names of the other six months, which expressed merely

the time, as is said, either of Numa or of one of the Tarquins) to 355 days, distributed over twelve months. It, however, soon appeared that the year of 355 days did not remain in harmony with the revolutions of the seasons, but that these were yearly becoming about ten or twelve days later, according to the calendar, than they had been the year before. The expedient adopted by way of remedy was to introduce into every second year a thirteenth month, called Mercedonius, consisting alternately of 22 and 23 days, and inserted immediately after the 23rd February. Thus a series of four years consisted of—

their numerical order, were *Quintilis* (fifth month), *Sextilis* (the sixth), *September* (the seventh), *October* (the eighth), *November* (the ninth), and *December* (the tenth). The two months afterwards added received the names of *Januarius* and *Februarius*,—the former in honor of the old Italian god, Janus, and the latter in honor of Februus, an Etrurian god, who seems to be identified with Pluto. Quintilis subsequently received the name of *Julius*, in honor of Julius Cæsar, and Sextilis the name of *Augustus*, in honor of that Emperor. It was afterwards attempted to change the names of September, October, November, and December into names derived from the Emperors Tiberius, Claudius, Nero, and Domitian; but the new names would not stick, and so the old remained. The months were divided by Calends (whence we have the familiar word Calendar), Nones, and Ides, which originally corresponded, the first to the new moon, the second to the first quarter, and the third to the full moon. This correspondence, however, was lost sight of after the solar year came into use, though the *names* were retained as convenient for making subdivisions, just as the *word month* is still retained in modern languages, although the *thing* signified has ceased to have any connection with a revolution of the *moon*.

	Days.
1. An ordinary year of 355 days,	355
2. A year of 355 days+Mercedonius, 22 days,	377
3. An ordinary year of 355 days,	355
4. A year of 355 days+Mercedonius, 23 days,	378
Total,	1465

But the true length of four solar years being only 1461 days, four Roman years as thus established would be four days too long; so that every four years the seasons would fall four days earlier in the year than they had done four years before, and in the short period of thirty years they would severally be moved back a month. This consequence becoming apparent, a remedy was sought by confiding to the pontiffs a discretionary power of intercalating as many days as they might consider necessary to preserve harmony betwixt the year of the calendar and the year of the seasons. But the pontiffs, abusing the power confided to them, their proceedings created so much disorder that at last the various festivals fell entirely out of coincidence with the seasons of the year at which they ought to have been celebrated.

§ 11.

It was reserved for Julius Cæsar, in his capacity of chief pontiff, not only to put an end to this confusion and the abuses in which it originated, but to establish a system of recording time which has descended to the present day, and is denominated, from its founder, the

Days.
.... 355
22 days, 377
.... 355
23 days, 378
————
.... 1465

rs being only
ablished would
four years the
the year than
he short period
moved back a
arent, a reme-
s a discretion-
as they might
y betwixt the
seasons. But
to them, their
that at last
if coincidence
they ought to

is capacity of
this confusion
it to establish
cended to the
s founder, the

Julian Calendar. He was aided in this great reformation by two eminent astronomers, viz., Marcus Flavius, a Roman, and Sosigines, an Egyptian—he himself having also paid some attention to astronomy when in Egypt, where the science was so far advanced that the length of the solar year was known to be about 365¼ days. The adoption of a civil year of exactly that duration, would, however, have involved inconvenient consequences. Thus if we suppose such a year to commence at midnight between 31st December and 1st January, the succeeding year would commence at 6 a.m. on the next 1st January; the next at noon on the following 1st January; the next at 6 p. m. on the 1st January of the third year; and, in fine, the next at the midnight between the 1st and 2nd January of the fourth year. Thus, in a series of four years, the first of January would be transferred piecemeal, quarter by quarter, backwards to the preceding year. Instead of this, Cæsar decided to adhere to years consisting of the whole number of days (365) and to allow the fraction of ¼ to accumulate until, at the end of four years, it should amount to an entire day, which was then to be added as a supplementary day to the year of its occurrence. The additional day given to the fourth year (which would thus consist of 366 days) was introduced where formerly the intercalary month Mercedonius had been introduced, namely, after the 23rd February, so that the 24th of February, (called in the Roman calendar a. d. VI. Kal. Martias) being reckoned twice over, received the name ante diem *bissextum* Kal. Mart., whence we have the name of *annus bissextilis* (bissextile) for the intercalary year.*

The future being thus provided for, it now remained to repair the consequences of past disorder, by bringing back the months to their normal position with respect to the seasons. To accomplish this, it was decreed that the year 708 from the building of the city (corresponding to the year B. C. 46, and to which chronologists have given the appellation of the *year of confusion*) should consist exceptionally of 445 days, by adding to the common year of 355 days, not only the intercalary month Mercedonius, but two extraordinary months, the first of 33 and the second of 34 days, which were inserted between the months of November and December. By this means the day of the vernal equinox was brought back to the 25th of March, the date which it was supposed to have held in the time of Numa Pompilius. A rectification of the intercalary system, which had fallen into some confusion after Cæsar's death, was effected by Augustus, in the year 8 B. C., on which occasion it was that he gave his own name to the month Sextilis. (§ 10.)

§ 12.

The Julian Calendar, together with the Jewish week of seven days (§ 9), was adopted by the Christian

* The common name given to bissextile years in our language is leap-year,—a term of which it is difficult to see the propriety. In the ecclesiastical calendars of the European continent, the day called "intercalary" in bissextile years is not the 29th but the 24th of February.

he intercalary

now remain-
disorder, by
position with
iis, it was de-
ig of the city
nd to which
of the *year of*
445 days, by
not only the
xtraordinary
days, which
ovember and
vernal equi-
rch, the date
the time of
intercalary
fusion after
n the year 8
ive his own

ewish week
e Christian

our language
he propriety.
ontinent, the
the 29th but

nations of Europe, and remained the common calendar of Christendom down to the year of our Lord 1582. In various states it continued to be used much longer, and in Russia it remains in use down to the present day.

§ 13.

It has been already stated that the interval of 365¼ days assumed in the Julian reformation as the length of a year, differs from its exact length by a fraction—though only a small fraction—of a day. As we have now to explain the part which this small fraction has played in chronology, it is requisite to attain and bear in mind a clear idea of the precise meaning of the term *year*.

A year, it has been said (§ 10), is the period after which the seasons are reproduced, or from the moment at which spring begins to the moment at which winter ends and spring recommences. This moment—and be it observed that we have here to do, not with days or hours, but with such minute periods of time as minutes, seconds, and fractions of a second—is considered to be the moment at which the centre of the sun's disc has such a position in the heavens that, if it were stationary there, day and night would be exactly equal—the sun would, during an apparent revolution of the heavens, be exactly twelve hours above and twelve hours below the horizon. But the sun's disc is not stationary. It has a continual easterly motion upon the heavens, moving at the rate of about 1° per day or 2½′ per hour, so that it does not retain the position in question more than a single second. It moves round the heavens as the hand of a clock moves round its dial, passing incessantly from point to point. The exact point at which the centre of the sun is at the moment above described, is therefore called the equinoctial point, as the moment of time at which it passes through that point is called the equinox. There are two equinoxes and two equinoctial points. The first, called the vernal equinox, because it has been agreed to fix on it as the epoch of the beginning of spring, takes place about the 21st March; and the second, called the autumnal equinox, because it has been agreed to fix on it as the epoch of the beginning of autumn, takes place about the 23rd September. The two equinoctial points are situate at opposite sides of the heavens, separated one from the other by an entire hemisphere.

If these two points maintained a fixed position on the heavens, the interval between the moments at which the centre of the sun's disc would pass twice successively through either of them, would be in fact the interval during which the sun makes, or appears to make, a complete revolution of the heavens. This interval is called the *siderial year*, and its exact length has been ascertained by astronomers to be 365 days 6 hours 9 minutes and $10\frac{38}{100}$ seconds.

But the equinoctial points have not a fixed position on the heavens. They are subject to a slow displacement from year to year in a direction contrary to the motion of the sun. The amount of this annual displacement is small, being a little less than the sixteenth part of a degree : but, small as the displacement is, it has been very precisely measured; and its effects, which

round its dial,
t. The exact
it the moment
inoctial point,
through that
wo equinoxes
ied the vernal
t on it as the
place about
he antumnal
on it as the
place about
l points are
eparated one

ition on the
ts at which
rice succes-
in fact the
appears to
rens. This
tact length
365 days 6

d position
r displace-
ary to the
displace-
eenth part
is, it has
is, which

are of the highest importance, as well in chronology as in astronomy, have been exactly appreciated.

On account of this removal of the equinoctial point backward, the sun, on making a revolution of the heavens, arrives at it sooner than it would have done had it not been displaced. This must be evident when it is considered that the equinoctial point, displaced in a direction contrary to that of the sun's motion, advances to meet the sun on its return. The sun therefore arrives at it before it makes a *complete* revolution of the heavens, and the time of each successive equinox *precedes* the time at which it would have taken place if the equinoctial point had been stationary. This phenomena has therefore been called the *precession of the equinoxes*. Its effect is, that the interval between two successive equinoxes, called the *equinoctial* or *tropical year*, is somewhat less than the *siderial* year.

The siderial year is of invariable length, and would on that account be well suited to be a standard measure of time; but it is rendered totally unfit for civil purposes by the circumstance that it is not in accordance with the periodic returns of the seasons. If the equinoctial points were stationary, the siderial year would also be the equinoctial year, and in that case it would be coincident with the return of the seasons. But in consequence of the displacement of the equinoctial points, the commencement of the equinoctial anticipates that of the siderial year; and although the extent of this anticipation is very small year by year, yet, accumulating during a long series of years, it causes the seasons to take place successively at every imaginable part of the siderial year.

If the precession of the equinoxes, or, in other words, the annual displacement of the equinoctial point, were regular and constant, the equinoctial year, differing from the siderial year by an inconsiderable quantity, would itself be invariable; and, as it is in accordance with the succession of the seasons, it would be in all respects eligible as a standard measure of civil time; but it so happens that, through the operation of several causes, the displacement is rendered variable; and though its variations are circumscribed within narrow limits, alternately increasing and decreasing, still, the existence of variation at all, renders the equinoctial year unfit for the purpose in question. A remedy for the variations has been found in the exact ascertainment of the *mean* annual displacement of the equinoctial point. This being done, a fictitious equinoctial point is supposed to exist, having this mean annual displacement; and, the interval between two successive returns of the sun to this fictitious equinoctial point being invariable, *that* interval is adopted as the standard, and is called the *mean solar or civil year*. Although it does not correspond rigorously with the returns of the seasons, it never varies from them by an interval great enough to be perceived or appreciated by any but astronomers: its exact length is 365d. 5h. 48m. 49$\frac{54}{1000}$s., being less than the siderial year by 20m. 20$\frac{3}{10}$s.

§ 14.

The exact length of a year that would always remain in accordance with the successive returns of the seasons being 365d. 5h. 48m. 49$\frac{54}{1000}$s., whilst the length

er words,
oint, were
differing
quantity,
a accord-
would be
e of civil
eration of
variable ;
d within
creasing,
the equi-
tion. A
he exact
nt of the
ous equi-
his mean
reen two
equinoc-
lopted as
ivil year.
with the
em by an
preciated
65d. 5h.
by 20m.

s remain
the sea-
e length

of the Julian year was 365d. 6h., it follows that the Julian year would depart from the course of the seasons at the rate of 11m. $10\frac{46}{1000}$s. per annum. The departure accumulating from year to year would amount to a whole day in 129 years, to two whole days in 258 years, to three in 387 years, and so on ; and although it would not be perceptible during the lives of a single generation, yet it must evidently become so after the lapse of centuries. The equinox falling back towards the beginning of the year at the rate of a day in 129 years was thus, in the fifteenth century, thrown back as much as eleven days, and would in the course of centuries have continually receded farther and farther, until it, and consequently the seasons, had successively assumed every possible position in the year. This fact, once ascertained, would of itself have afforded a sufficient reason for undertaking a revision and correction of the calendar. The immediate cause of the revision which took place was as follows :

§ 15.

It was the rule of the church to celebrate the festival of Easter at a time not far removed from the 21st March, which was taken to be the day of the equinox, depending, however, also upon conditions connected with lunar phenomena with which we are not at present concerned. If, therefore, the real equinox were continually falling back (§ 14) farther and farther from the 21st March, it must obviously happen that the festival of Easter, still related to the 21st March, would, in the course of time, fall successively in every season of the year. This error began to be fully recognised about the beginning of the sixteenth century; and after various attempts had been made to excite attention to the subject, it was taken up in earnest in the year 1577, by Pope Gregory XIII., under whose auspices a reformed calendar was prepared, which received the name of the *Gregorian calendar*. Gregory entrusted the superintendence of the whole reformation to Christopher Schlüssel, better known by his latinized name of Clavius, (+1712 æt. 75) a learned Jesuit, well known to mathematicians as a commentator on Euclid and other ancient geometers; and the changes introduced, which were sanctioned by a bull dated 24th February and published 1st March, 1582, were as follows:

As at the epoch of the Julian reform, two errors were to be corrected, those of the past and those of the future. The accumulated effect of the past errors was, that the nominal epoch of the vernal equinox (21st March, the date on which it fell at the time of the Council of Nice in 325) had fallen ten days behind the real date of its occurrence. To remedy this, it was decreed that ten nominal days should be destroyed—an expedient which has since been by some considered unnecessary, and which undoubtedly has given rise to confusion and chronological error. The day which should have been the 5th of October, 1582, was ordered to be called the 15th, so that the ten days from the 5th to the 14th of October, inclusive, were never allowed to exist. The last ten days of 1582, according to the Julian Calendar, were thus thrown over into 1583, in-

recognised
; and after
ttention to
year 1577,
ices a re-
d the name
rusted the
o Christo-
d name of
ell known
uclid and
es intro-
ated 24th
re as fol-

rors were
of the fu-
rors was,
ox (21st
e of the
ehind the
s, it was
stroyed—
onsidered
n rise to
y which
ordered
a the 5th
allowed
g to the
1583, in-

asmuch as the 21st of December, 1582, became the 31st of December, 1582, and consequently 22nd December, 1582, became 1st January, 1583. The 11th March, 1583, according to the Julian Calendar, upon which the equinox really fell, thus became the 21st of March in the Gregorian Calendar, and the day of the equinox was secured for the future in the undisturbed possession of that date by the following arrangement:

We have already seen (§ 14) that the Julian year is too long by 11 minutes $10\frac{48}{100}$ seconds. This will be found to make 400 years too long by about 3 days; so that the error may be nearly corrected by reducing 3 leap-years in 400 years to common years. The mode of doing this was by neglecting to make leap-years of the *anni domini* which end with 00, unless the preceding figures are divisible by 4. Thus 1600 is leap-year, but 1700, 1800, 1900 are not leap-years. In the Julian Calendar *every* year is leap-year which is divisible by 4; in the Gregorian Calendar, the exception is that the *secular* years (those closing centuries, or ending with 00) are not leap-years unless they are divisible by 400. This correction leaves the year still a little too long. The difference would amount to a day in about 3600 years. To provide for this difference, the French astronomer, Jean Joseph Delambre (born 1749,+1822) proposed, and the proposal was adopted in the French revolutionary Calendar,* that the years of the French

* This Calendar was introduced by a decree of the National Convention, dated 5th October, 1793, establishing the Era of the French Republic, the epoch, or point of departure, of which was the midnight at which commenced the day of the autumnal

Republic 3600, 7200, and 10,800 should not be leap-years. But Pope Gregory did not aspire to legislate for so remote a posterity.

§ 16.

Notwithstanding the undeniable reasonableness of Gregory's reform, the adoption of the change was far from universal. Protestant states and the Greek Church were equally disinclined to it, because it ema-

---

equinox (22nd September) 1792. Each subsequent year was in like manner to commence at midnight preceding the day of the true autumnal equinox. The year consisted of twelve months of 30 days each, to which were added in ordinary years 5, in leap-years 6 complementary days. Instead of weeks, the month was divided into three decades, or periods of ten days. The months received names denoting some characteristic quality. The autumnal months (22d September to 20th December) were *Vendimiaire* (Vintage month), *Brumaire* (Fog month), and *Frimaire* (Hoar-frost month); the winter months (21st December to 20th March) were *Nivose* (Snow month), *Ventose* (Wind month), and *Pluviose* (Rain month); the spring months (21st March to 18th June) were *Germinal* (Bud month), *Floreal* (Flower month), and *Prairial* (Meadow month); and the summer months (19th June to 17th September) were *Messidor* (Harvest month), *Thermidor* (Heat month), and *Fructidor* (Fruit month). The last was followed by the *jours complémentaires*, of which the first (17th September) was called *la fête du génie;* the second, *la fête du travail;* the third, *la fête des actions;* the fourth *la fête des récompenses;* and the fifth (21st September) *la fête de l'opinion.* The ten days of each decade were called *Primidi, Duodi, Tridi, Quartidi, Quintidi, Sextidi, Septidi, Octidi, Nonidi,* and *Decadi.* The last named was the day of rest. At the instance of Bonaparte, by a decree of the Senate, dated 9th September, 1805, to take effect from 1st January, 1806, the revolutionary calendar was abolished and the Gregorian restored.

not be leap-
to legislate

lableness of
nge was far
the Greek
ause it ema-

year was in
he day of the
welve months
ars 5, in leap-
le month was
The months
lity. The au-
were *Vendi-*
nd *Frimaire*
mber to 20th
month), and
[arch to 18th
wer month),
months (19th
onth), *Ther-*
]. The last
uch the first
cond, *la fête*
urth *la fête*
) *la fête de*
led *Primidi,*
idi, Octidi,
of rest. At
te, dated 9th
1806, the re-
ian restored.

nated from papal authority: truth was no longer truth
when it issued from the Vatican ; or, as it was wittily
expressed, they preferred being in opposition to the sun
to being in harmony with the Pope. The commencement of the reform was fixed by the papal bull for the
day following the feast of St. Francis, that is to say,
the 5th October, 1582, which was decreed to be called
the 15th October; and the change was then adopted
in Italy, Spain, and Portugal. It was adopted in France
on 10th December of the same year, which was called
20th December. It was adopted in 1584 in the Catholic
states of Germany, the Catholic cantons of Switzerland,
and the Catholic provinces of the Netherlands. In Poland
it was adopted in 1586; in Hungary in 1587. The
Protestant states of Germany, having resisted the
reform for nearly 120 years, yielded at length : they
accepted it in 1700, of which year the 19th February
was declared to be the 1st March. Denmark and the
United Provinces made the change simultaneously with
Protestant Germany, and their example was followed
the subsequent year by the Protestant cantons of Switzerland, which commenced the eighteenth century with
12th January, 1701. In England, where the ante-
papal spirit was probably stronger than anywhere else,
the reform was resisted for nearly two centuries, so
that the real date of the equinox had come to fall
eleven days earlier than the legal one. In 1752, however, the reform was adopted in England,—the 3d being declared to be the 14th of September. The change
excited an immense deal of dissatisfaction, chiefly, no
doubt, amongst the lower orders, but by no means confined

exclusively to that portion of the community. The mob pursued the minister in his carriage, clamouring for the eleven days by which, as they supposed, their lives had been shortened ; and the illness and death, which shortly afterwards occurred, of the astronomer Bradley, who had assisted the government with his advice, were attributed by many to a judgment from heaven.* The last country that adopted the change was Sweden, which in 1753 reckoned the day following 17th February as 1st March. The Russians, and in general, the schismatic Greek Church, still adhere to the Julian calendar ; and accordingly, by the further accumulation of the effects of the erroneous length assigned to the year, the Russian legal equinoxes are now (since 1800) twelve days behind the real equinoxes. In the year 1900, the difference will amount to 13, and in 2100 to 14 days.

§ 17.

At present, throughout Christendom, the civil year commences from midnight of 31st December, that is, with 1st January ; but this uniformity of reckoning is of modern and almost recent date. Amongst the French, in the time of Charlemagne, the year commen-

---

* The great promoters of the change in England, and the parties by whom the bill was introduced into the House of Peers, were the Earl of Macclesfield and the Earl of Chesterfield, both pupils of the celebrated mathematician DeMoivre. Lord Chesterfield alludes to his hand in it once or twice in his letters to his son. The last survivor of those whom they consulted was Charles Walmesley (born 1721,+1797), a Benedictine monk, afterwards vicar apostolic of the western district of England, well known as a mathematician, and a member of the Royal Society.

nity. The
lamouring
osed, their
and death,
astronomer
t with his
ment from
the change
y following
ns, and in
adhere to
the further
length as-
tes are now
equinoxes.
to 13, and

e civil year
er, that is,
eckoning is
mongst the
ar commen-

, and the par-
use of Peers,
iterfield, both
, Lord Ches-
letters to his
onsulted was
ine monk, af-
England, well
Royal Society.

ced on Christmas-day : amongst the same poeple, under the monarchs of the house of Capet, it commenced on Easter-day ; and this was a very general date in the twelfth and thirteenth centuries : in England down to the change of style in 1752, it commenced on Lady-day (25th March).

§ 18.

A matter of more importance than the day of the year's commencement is the mode of computing by years, so as to be able to refer historical events to distinct times as the times of their occurrence. This is most conveniently done by selecting some remarkable event to which the other events of history are to be referred as having occurred so many years before or so many years after it. Such an event is termed an *epoch*, and the period of which it forms the commencement or point of departure is termed an *era*. The number of epochs which have been used by different nations, and by the same nation at different periods or for different purposes, is considerable. The Jews employed as epochs, the Exodus, the Building of the Temple, the Accession of Herod, the Destruction of the Temple; and the epoch from which they now date is the Creation of the World, which they reckon to have taken place 3761 years before the Christian epoch. The Greeks used to count from Cecrops, from the destruction of Troy, from the official years of the Ephori in Sparta and of the Archons in Athens. About 300 years before Christ, Timeus, a Sicilian historian, introduced the practice of reckoning by *Olympiads*, or the quadrennial recurrence of the Olympic games. These games were celebrated once

c

every four years between the new and full moon first
following the summer solstice, on the banks of the
river Alpheus, near the city of Pisa in the Peloponne-
sus, and lasted five days.   Their original foundation is
obscure, and has been connected with legendary tales
about Pelops and Herakles.   After having been ne-
glected during several centuries, they are said to have
been revived by Iphitus, King of a canton of Elis, in con-
cert with Lycurgus, the lawgiver of Sparta.   This much
is certain, that they were finally established in the year
B.C. 776, when the practice was introduced of inscrib-
ing in the gymnasium of Olympia the name of the vic-
tor in the contests.   The first who received this honor
was Coroebus ; and the games in which he was victorious were celebrated about the time of the summer
solstice in the year just mentioned, which accordingly
is the first year of the 1st Olympiad, and also the 3938th
year of the Julian period (See. § 21), and 23 years before
the foundation of Rome, according to Varro.   The
293d and last Olympiad occurred A.D. 394, after which
the games themselves were suppressed as heathenish by
the Emperor Theodosius.   The Romans used to reckon
from the building of the city (A. U. C.), from the years
of the consuls, and from the years of the Emperors.
The consulship having become extinct in the reign of
Justinian (A. D. 541), the mode of computing by consular
years then necessarily ceased for the future.   After the
sixth General Council (A.D. 681), the prevailing epoch
amongst the Greeks was the Creation of the World,
which, following the Septuagint version of the Old Testament, they placed 5508 years before the Christian

moon first
ks of the
'eloponne-
ndation is
idary tales
; been ne-
iid to have
llis, in con-
This much
in the year
of inscrib-
of the vic-
this honor
vas victori-
he summer
iccordingly
the 3u38th
years before
arro. The
after which
iathenish by
d to reckon
m the years
e Emperors·
the reign of
; by consular
. After the
ailing epoch
the World,
the Old Tes-
he Christian

epoch. This era, called the era of Constantinople, was long in general use throughout the Eastern Empire, and continued to be employed in Russia for both civil and ecclesiastical purposes down to the year 1700, when the Christian era was introduced by Peter the Great. In the Western Empire, the era of Diocletian, commencing with the accession of that Emperor (29th August, A.D. 284), and called also, from the ten years' persecution of the Christians which occurred in that reign, the era of the Martyrs, seems to have extensively prevailed. In Egypt it continued to be the civil era down to the conquest of that country by the Arabians; and, down even to the present day, it continues to be used in the ecclesiastical calendar of the Coptic and Ethiopian Christians. In the sixth century, the Abbot Dionysius, surnamed (by himself, from modesty) Exiguus, or the little, a Scythian by birth, but who died in Rome about the year 556, proposed the year of our Saviour's birth as a chronological epoch; and this epoch, giving rise to the Christian era, has since come to be generally adopted amongst Christian nations. A proposal exceedingly similar had already been made (A.D. 465) by another churchman, Victorinus or Victorius of Aquitaine,—the chief difference being, that Victorinus had proposed dating from the year of Christ's crucifixion, not of his birth. The latter event (his birth) Dionysius was led to conclude had occurred on the 25th December, in the 753d year from the foundation of Rome; and, according to the mode of reckoning finally adopted, the year 1 A.D. was that which commenced at midnight between 31st December of the 753d, and 1st January of the 754th year of Rome,—that being the year in which Christ was supposed to have completed the first year of his age. In fixing upon that year, however, it is obvious that Dionysius has fallen into error. It appears from the narrative of Saint Matthew (c. ii. v. 1–19) that our Saviour was born towards the latter end of the reign of Herod the Great; and as Herod died, according *o Josephus, in the year of Rome 750, at which time Jesus was with his parents in Egypt, his birth must have taken place at least four, and probably six years earlier than Dionysius computed. This circumstance, however, does not in any respect affect the value of the epoch for chronological purposes; because, whatever may have been the actual year of our Saviour's birth, it is the first day (1st January) of the year of Rome 754, that connects the Christian with the ancient chronology. In order, therefore, to convert any year of the Christian era into the corresponding year of Rome, it is only necessary to add 753 to it. Thus the year 1 A.D. was the year 754 of Rome; the year 20 A.D. the year 773 of Rome, and so on.

§ 19.

Since, according to the Christian chronology, time is counted thus prospectively forward from the birth of Christ, the *year after* that event being taken as the first year of the series, it might by analogy be presumed that, in counting time retrospectively, the *year before* the same event would be taken as the first year of the backward series. Thus, while the year after the birth of Christ is 1 A.D., the year before that of the birth of

 year in which
the first year of
rever, it is ob-
or. It appears
(c. ii. v. 1–19)
tter end of the
d died, accord-
750, at which
zypt, his birth
and probably
ted. This cir-
respect affect
ical purposes;
actual year of
st January) of
Christian with
'ore, to convert
corresponding
add 753 to it.
of Rome; the
so on.

ronology, time
from the birth
g taken as the
ry be presumed
the *year before*
st year of the
after the birth
of the birth of

Christ would be the year 1 B.C., and consequently the year itself in which Christ was born would be either 0 A.D. or 0 B.C. indifferently. By such a mode of expressing dates, the interval between any day in any year A.D. and the corresponding day in another year B.C. would be found by adding together the numbers expressing the years. Thus the interval between 1st July 1 A.D. and 1st July 0 B.C. was 1 year; the interval between 1st July 1 A.D. and 1st July 1 B C. was 2 years; the interval between 1st July 15 A.D. and 1st July 14 B.C. was 29 years, and so on. And this is, accordingly, the mode of expressing dates which astronomers employ. Unfortunately, however, it is not the mode which has been adopted by historians and chronologists. According to these, the year 753 of Rome, in which Christ is supposed to have been born, is the year 1 B.C., and consequently all their dates B.C. exceed the corresponding dates of astronomers by 1. Thus the year which astronomers call 500 B.C., historians call 501 B.C. To find, therefore, the interval between any day in a year A.D. and the corresponding day in any year B.C. when the historical dates are used, it will be necessary to add together the two dates and subtract 1 from their sum.

§ 20.

Historical events are often referred to as having occurred in a particular *century*. A century is a complete period of a hundred years. Eighteen such periods have already elapsed since the commencement of the Christian era, and we are now living in the nineteenth. The first century commenced at the instant of midnight preceding the 1st January 1 A.D. and ended at midnight of 31st December 100 A.D. The eighteenth century commenced with 1st January 1701, and ended with 31st December 1800. In like manner, the first day of the nineteenth century was 1st January 1801, and the last will be 31st December 1900.

§ 21.

Down to the latter part of the sixteenth century, chronology, as a science, can scarcely be said to have existed. All ancient history had been written in a servile and uncritical spirit, copying dates as it did everything else from the authorities immediately under the compiler's eye, with little or no endeavour to reconcile discrepancies, or to point out any principles of computation. In 1583, however, Joseph Scaliger, perhaps the most learned man either of his own or of any other age, published a work on the rectification of dates (*De emendatione temporum*), to which nothing that the age produced was superior, or perhaps equal, in originality, depth of erudition, or vigorous encountering of difficulties. It was a work requiring, besides much attention and acuteness, an amount of erudition, of which he alone in Europe could be reckoned master. He has been censured for introducing too many conjectures, and drawing too many inferences from conjectural data. But, whatever may be his merits in the determination of particular dates, he is certainly the first who laid the foundation of chronology as a science. It is to him we owe the invention of what is termed the *Julian period*, which was meant as an era of common reference to

and ended at
The eighteenth
701, and ended
nner, the first
January 1801,

.

eenth century,
e said to have
written in a
ates as it did
ediately under
vour to recon-
principles of
Scaliger, per-
own or of any
ation of dates
thing that the
qual, in origi-
countering of
ides much at-
tion, of which
ster. He has
conjectures,
jectural data.
determination
first who laid
It is to him
Julian period,
reference to

which dates referred to other eras might be reduced, so as to form a common standard of historical and chronological time. This he found might be effected by taking a cycle of 7980 years, of which he fixed the commencement at the moment of noon for the meridian of Alexandria on 1st January 4713 B.C., and by the employment of this period, according to the testimony of the best modern chronologists, light and order were for the first time let in upon the obscurity and confusion in which ancient history and chronology were involved.*

§ 22.

About twenty years after Scaliger's death, Denis Petau or Petavius, a Jesuit (born 1583, died 1652), with learning scarcely inferior to Scaliger's own, published a very valuable work on the same subject. Using greater caution than his predecessor, and dissenting from many of his conclusions, he assailed him rather rudely, and displayed too great anxiety to refute him whether right or wrong. But his work was clear and methodical, and long served as a text-book in the universities. The first important contribution to the science made by an English writer, was Archbishop

* Since the year of the birth of Christ was the 4713th of the Julian period, the order of any later year of the Christian era in the Julian period will be found by adding 4713 to the year. Thus for example, the year 1859 is the 1859+4713=6572d year of the current Julian period. To find the order of any year before Christ in the Julian period, it will only be necessary to subtract the year from the order of the year 1 of the Julian period, that is from 4714. Thus, the date of the battle of Cannæ being 216 B.C., its date in the Julian period is 4714—216=4498.

Usher's Annals of the Old Testament, the first part of which was published in 1650, and the second part followed a few years afterwards. Usher's chronology is that which has been generally adopted by English historians; and, having also been followed by the continental writers Bossuet, Calmet, and Rollin, it formed for many years the common chronological scheme of Europe. In our own day the chronologist of most weight and reputation is probably Christian Ludwig Ideler, Professor in the University of Berlin, from whose carefully elaborated works, from those of the other writers just named, and from a learned work, *L'art de vérifier les dates*, published by the Benedictines of St. Maur, the more popular recent works on chronology have mostly been compiled.

§ 23.

To the study of History, the science of Chronology forms a necessary and one of its most important auxiliaries. It is true, that, in spite of the many laborious inquiries that have been instituted and pursued within the last three centuries, the chronology of very remote events remains obscure and conjectural; and the most judicious inquirers have of late been coming to the opinion, that, with some few exceptions, there are no means of establishing accurate dates before the Olympiads. But as it is impossible to attain a comprehensive view of historical events without marshalling them in some arrangement as regards their coincidence and sequence in time, we are forced from the very outset of our historical studies to recur, at least provisionally, to

e first part of
:cond part fol-
chronology is
y English his-
by the conti-
lin, it formed
cal scheme of
gist of most
stian Ludwig
a, from whose
of the other
work, *L'art*
nedictines of
a chronology

Chronology
ortant aux-
ny laborious
:sued within
very remote
nd the most
ning to the
iere are no
the Olym-
comprehen-
alling them
idence and
y outset of
ionally, to

some chronological system, even though we may utterly distrust its near approach to correctness. As we descend the stream, our chronology becomes gradually more reliable, until at last the date of every important occurrence may be ascertained with precision.

The value of the following tabular exhibition of the more prominent events of universal history will probably consist less in the aid which it may afford to the pupil's imagination in combining these events into connected groups, than in the assistance which it will afford the memory, in recalling to it, by means of a word or a sentence, the substance of a lesson. The divisions and subdivisions into greater and lesser historical periods are those which are most usually adopted for the purpose of tuition. The three great periods of Ancient, Medieval, and Modern History, although they imply no interruption of continuity in the march of events, but on the contrary are linked closely together by multifarious ties, are yet characterized by features so essentially different, and by peculiarities and contrasts so striking, that each of them may in a certain sense be regarded as a separate and independent whole. An attempt has been made to indicate in a few sentences the general features of each; but any thing like a nearer consideration of these would lead us far beyond the limits of the present outline.

ANCIENT HISTORY.

B.C. 4000—A.D. 476.

Ancient History, embracing the period from the creation of man to the overthrow of the Roman Empire of the West, contains *first*, the primitive history of the human race from Adam to Noah, and the division of peoples which took place shortly after the Flood; *secondly*, the fabulous history, or mythological period, of the earliest nations; *thirdly*, the historical period, or proper history, of the earliest nations of Asia, and of the settlers along the shores of the Mediterranean Sea.

FIRST PERIOD.

B.C. 4000—B.C. 555.

FROM ADAM TO CYRUS.

Primitive history of the human race from Adam to Noah.—Diffusion of Noah's descendants over the earth.—Foundation of the first great empires in Assyria and Babylonia.—Overthrow of the New Babylonian Empire by the Medo-Persians.—During this period Asia is the principal scene of historical events.

B. C.

4000. Adam and Eve in Paradise. Central Asia the original abode of the human race. Cain, Abel, Seth.

B.C.

2300. Noah, the second universal parent of the human race. The deluge. Noah's sons, Shem, Ham, and Japheth.

2000. The Patriarchs, Abraham, Isaac, Jacob. The first great states, Babylonia and Assyria (Babylon and Niniveh). Ninus and Semiramis. Ancient civilization of Egypt. Memphis, Thebes.

1900. Joseph in Egypt. The Phenicians the most ancient of trading nations. Their cities, Tyre and Sidon. Foundation of Colonies. Invention of letters, glass, purple. Emigration of the Pelasgians from Asia Minor into Greece.

1600. Athens founded by Cecrops from Egypt; Thebes by Cadmus from Phenicia. Danaus from Egypt in Argos. The Greeks, or "Hellenes," in Hellas.

1500. Moses in Egypt. Exodus of the Israelites. The law given from Mount Sinai. Conquest of Palestine. The Judges. Sesostris in Egypt.

1300. Immigrations into Greece from Egypt and Phenicia. Pelops.

1250. Argonautic expedition. Hellenic heroes: Jason, Hercules, Theseus, Œdipus. The musicians Orpheus and Amphion.

1200. Trojan war. Destruction of Troy. Grecian heroes: Agamemnon, Menelaus, Achilles, Ajax, Ulysses. Trojan heroes: Priam, Hector, Paris (the husband of Helen). Æneas escapes to Italy. Ascanius builds Alba Longa. The Latins; the Etruscans.

B.C.

1100. Samuel, the last of the Israelitish judges. Saul, the first of their kings. His successor, David. Jerusalem. Codrus, the last king of Athens. Greek colonists settle in Asia Minor, Sicily, and Lower Italy.

1000 Solomon, under whom the Hebrew monarchy attains its highest splendour. Its division, on his death (975), into the kingdom of Judah, under Rehoboam, and the kingdom of Israel, under Jeroboam. Homer.

888. Lycurgus, the Spartan lawgiver. Carthage, in Africa, founded by Dido with a Phenician colony. Sardanapalus, the last ruler of the Assyrian empire. Media forms an independent state.

776. Commencement of the Hellenic mode of computation by olympiads, or periods of four years, so named from the games celebrated once in four years at Olympia, in Elis. Delphic oracle.

754. Rome founded by Romulus and Remus. 1. Romulus first king. 2. The peaceful Numa Pompilius. 3. The warlike Tullus Hostilius conquers Alba Longa. 4. Ancus Martius. 5. The wise Tarquinius Priscus. 6. Servius Tullius.

722. The kingdom of Israel, with Samaria the capital, destroyed (in the reign of Hosea) by Shalmaneser, king of Assyria. Isaiah prophecies in Judah. Destruction of Sennacherib's army before Jerusalem (715).

600. The New Babylonian empire founded on the ruins of the Assyrian by Nabopolassar. The city of

litish judges. Saul,
is successor, David.
ist king of Athens.
ia Minor, Sicily, and

lebrew monarchy at-
Its division, on his
om of Judah, under
m of Israel, under

river. Carthage, in
a Phenician colony.
of the Assyrian em-
endent state.
nic mode of compu-
ods of four years, so
rated once in four
Delphic oracle.
nd Remus. 1. Ro-
ceful Numa Pompi-
Hostilius conquers
tius. 5. The wise
ius Tullius.
Samaria the capital,
a) by Shalmaneser,
ophecies in Judah.
army before Jeru-

unded on the ruins
ssar. The city of

B.C.
600. Tyre destroyed by his son Nebuchadnezzar, a
contemporary of the Athenian legislator, Solon.
588. Destruction of the kingdom of Judah under its
last king, Zedekiah. The prophet Jeremiah.
Babylonian exile : The prophets Ezekiel and
Daniel.
560. Croesus, king of Lydia. Pisistratus, tyrant of
Athens. The seven wise men of Greece : Solon,
Thales, Periander, Kleobulos, Pittakos, Bias,
Cheilon.
Æsop the fabulist. Pythagoras the Philosopher.

SECOND PERIOD.

B.C. 555—B.C. 333.

FROM CYRUS TO ALEXANDER.

From the foundation of the second great empire to
its overthrow by the third,—the Græco-Macedonian.—
Rise of freedom and civilization in the West : their
struggle against the power of the East in the Græco-
Persian wars: Victory of Grecian culture, and its diffusion by means of Alexander's expedition.—Beginnings
of the Roman power in Italy.—Renewal of the Israelitic
state.—Western Asia and South-Eastern Europe form
the main scene of events.

B.C.
555. Cyrus, the founder of the Medo-Persian empire,
conquers Croesus, subdues Asia Minor and Ba-

B.C.

555. bylonia, and permits the Jews to return to Palestine (under Ezra and Nehemiah). Re-building of Jerusalem. Death of Cyrus (529). His son and successor Cambyses conquers Egypt, and reduces it to a satrapy of the Persian empire. Zoroaster, founder of the Persian religion (Zend-avesta). Confucius, founder of the Chinese religion.

509. Expulsion of Tarquinius Superbus, the seventh and last king of Rome. Brutus and Collatinus, the first consuls. Porsenna and the Etruscans. Revolt of the Greeks of Asia Minor against the Persians.

493. Occupation of the Sacred mount by the Roman plebeians. Menenius Agrippa. Tribunes of the people.

490. Coriolanus, banished from Rome, takes refuge amongst the Volsci, and becomes the leader of an expedition against his native city, which is saved by the intervention of his mother. First war of the Persians and Hellenes. Darius Hystaspis conquered at Marathon by Miltiades the Athenian.

480. Second Græco-Persian war: Xerxes: Heroic death of Leonidas with 300 Spartans at Thermopylæ. Naval victory of Themistocles at Salamis. Aristides the Just.

479. Mardonius defeated by Pausanias and Aristides at the battle of Platæa. Themistocles, banished from Athens, dies in Asia Minor. Cimon, the son

o Pales-
uilding
His son
pt, and
empire.
(Zend-
ese re-

eventh
atinus,
uscans.
ast the

Roman
of the

refuge
der of
hich is
First
s Hys-
es the

Ieroic
ermo-
lamis.

des at
ished
le son

B.C.

479. of Miltiades, defeats the Persians at Eurymedon
by sea and land. End of the Persian wars (466).

444. Pericles, orator and statesman, at the head of the
Athenian republic, now at the height of its pros-
perity and importance. Phidias the sculptor.
Herodotus, the father of history. The three
great poets Æschylos, Sophokles, and Euri-
pides. The painters Zeuxis and Parrhasius. In
Rome, the decemvirs: laws of the twelve tables:
Appius and Virginia.

431. Peloponnesian war in Greece betwixt Sparta, the
head of the aristocratic (Dorian), and Athens,
the head of the democratic (Ionian) states.
Plague at Athens, and death of Pericles (429).
Hippocrates.

404. End of the Peloponnesian war. Sparta is now
the most powerful of the Grecian states. War
betwixt Cyrus the younger, and his brother,
Artaxerxes. Socrates condemned to swallow
poison. His scholars Plato and Xenophon.

396. Agesilaus, king of Sparta, vanquishes the Persians in Asia Minor. Rome sacked by the Gauls
under Brennus. Camillus, the conqueror of Veii,
created dictator, and defeats the Gauls.

360. Philip II. king of Macedon. His Athenian oppo-
nent, the orator Demosthenes.

338. Battle of Chæronea. Philip master of Greece.
Roman wars with the Samnites, ending in the
subjection of the latter.

THIRD PERIOD.

B.C. 333—B.C. 30.

FROM ALEXANDER TO AUGUSTUS.

After a brief period of splendour, the Macedonian empire is rapidly dissolved. Meantime, the fourth great empire, the Roman, is gradually rising, until, mistress of the East as well as of the countries bordering on the Mediterranean, the history of the Roman state becomes the history of the known world.—Under the influence of internal corruption and violent political revolutions, the two triumvirates end in the establishment of the Empire.—Italy becomes the centre of the " orbis terrarum."

B.C.

333. Alexander the Great, son of Philip II. of Macedon, founds the great Græco-Macedonian empire; crosses the Hellespont at the head of 35,000 men; gains three great victories over Darius Codomannus; subdues the Persian empire, and penetrates to India. Foundation of Alexandria in Egypt. His contemporaries Apelles, Aristotle, Diogenes, Zeno, Epicurus.

323. Alexander dies at Babylon. Out of the ruins of his vast empire, various smaller states are founded by his generals :—Seleucus in Syria ; Ptolemy in Egypt. Translation of the Old Testament into Greek (Septuagint). Alexandria becomes the seat of Greek learning under the Ptolemies. Pyrrhus, king of Epirus, at war in

s.

Macedonian
, the fourth
rising, until,
ntries border-
f the Roman
rld.—Under
lent political
he establish-
entre of the

of Macedon,
ian empire;
d of 55,000
r Darius Co-
empire, and
Alexandria
lles, Aristo-

the ruins of
states are
s in Syria;
the Old Tes-
Alexandria
under the
s, at war in

B.C.

323. Italy with the Romans, who defeat him and sub-
due Tarentum, thereby establishing their dominion over all Lower Italy.

264 First Punic war (betwixt the Romans and the Car-
to   thaginians) arising principally out of disputes
241. about Sicily. First naval victory of the Romans,
under Duilius. Regulus in Africa, at first victorious, afterwards a prisoner.

218 Second Punic war. Setting out from Spain,
to   which had been subdued by his father, Hamilcar
201. Barcas, Hannibal, at the head of 59,000 men,
enters on his Italian expedition : he defeats the
Romans in four great battles, at the Ticinus, the
Trebia, the Lake of Trasimene, and Cannæ.
Fabius Maximus Cunctator, "the shield," and
Marcus Marcellus, "the sword" of the Romans.
Archimedes defends Syracuse against the latter.
Hannibal in Capua.

202. Battle of Zama, in Africa, where Hannibal is defeated by P. Cornelius Scipio Africanus. Peace
concluded following year.

200. Romans at war with Philip III. of Macedon : victory of Flaminius at Cynocephalæ. Roman war
in Syria against Antiochus III. (the Great).

183. Death of Hannibal and of Scipio Africanus. Roman writers Polybius, Ennius, Plautus, Terence.

168. Battle of Pydna, where Æmilius Paulus defeats
Perseus, king of Macedon. In Palestine the
Maccabees head a Jewish revolt against Anti-
ochus IV.

D

B.C.

149 Third and last Punic war. Carthage destroyed
to   by P. Cornelius Scipio Africanus the younger.
146. Africa a Roman province. Corinth destroyed
      by Mummius. Greece a Roman province.

133. Popular disturbances in Rome instigated by the
     two brothers Tiberius and Caius Gracchus.
     Commencement of permanent civil discords
     (Agrarian law).

111. War against Jugurtha, king of Numidia, conduct-
     ed, first by Metellus, afterwards by Marius and
     Sylla.

101. The Cimbri and Teutones defeated by Marius at
     Aquæ Sextiæ. Social war (90).

 88. Civil war betwixt Marius and Sylla. Defeat by
     the latter of Mithridates, king of Pontus. Ma-
     rius and Cinna in Rome. After their death
     Sylla's reign of terror. Complete reduction of
     Mithridates by Lucullus and Pompey. The ser-
     vile war, which had broken out in Italy under
     Spartacus, extinguished by Pompey and Crassus.

 63. Catiline's conspiracy: discovered and defeated
     by Marcus Tullius Cicero ("pater patriæ").

 60. First triumvirate, formed by Cn. Pompey, C. J.
     Cæsar, and M. L. Crassus.

 49. Outbreak of the civil war between Pompey and
     Cæsar: the latter crosses the Rubicon: Battle
     of Pharsalia (48): Pompey murdered in Egypt:
     Death of Cato the younger at Utica: Battle of
     Munda (45): Cæsar dictator: Aspires to the
     crown.

ge destroyed
the younger.
th destroyed
rovince.
igated by the
us Gracchus.
ivil discords

dia, conduct-
y Marius and

by Marius at

. Defeat by
Pontus. Ma-
their death
reduction of
ey. The ser-
a Italy under
and Crassus.
nd defeated
patrie ").
mpey, C. J.

Pompey and
icon : Battle
ed in Egypt :
:a : Battle of
pires to the.

B.C.
46. Julian Calendar : year of confusion.
44. Murder of Cæsar by Brutus, Cassius, and others.
43. Second triumvirate : M. Anthony, Octavian, and
    Lepidus. Battle of Philippi, and death of Bru-
    tus and Cassius.
31. Civil war between Anthony and Octavian. Battle
    of Actium. Anthony and Cleopatra die at Alex-
    andria. Octavian sole ruler, under the name of
    Cæsar Octavianus Augustus. End of the Roman
    Republic and commencement of the Empire.
    Rome now a city containing about two millions
    of inhabitants. Mæcenas a statesman and pa-
    tron of letters. Horace ; Virgil ; Ovid ; Corne-
    lius Nepos. Golden age of Roman literature.

FOURTH PERIOD.

B.C. 30—A.D. 476.

FROM AUGUSTUS TO ROMULUS AUGUSTULUS.

Embraces the history of the Roman Empire until its
division into the Empires of the West and the East (Rome
and Constantinople).—Conquest of Christianity over
Heathenism.—Decline of Roman power and civilization.
—The West is overpowered by the Teutonic nations,
and the East again becomes Greek.—Commencement
of the migrations from East to West.—Culture pro-
gresses towards the North.

A.D.

*1. Jesus Christ born at Bethlehem.

9. Varus defeated by Arminius (Hermann), leader of the Cheruscians.

14. Augustus dies, and is succeeded by his step-son, Tiberius.

33. Crucifixion of our Saviour.

37. Caligula. Claudius (41). Commencement of Roman conquests in Britain.

54. Nero : First persecution of the Christians. Galba. Otho. Vitellius.

70. Destruction of Jerusalem by Titus. Roman war in Britain.

79. Titus Emperor. Eruption of Vesuvius : destruction of Herculaneum and Pompeii, and death of the elder Pliny. Britain subdued by Agricola : his son-in-law Tacitus the historian.

100. Trajan : the Roman empire at its greatest extent. Pliny the younger : Plutarch.
Hadrian.
Antoninus Pius.
Marcus Aurelius Antoninus.

180. With Commodus begins a series of bad emperors, amongst whom is the infamous Heliogabalus, the priest of the sun. The Roman empire at its lowest pitch of internal corruption. A very few good emperors, as Septimius Severus, Alexander Severus, and Aurelian, the conqueror of Zenobia of Palmyra.

---

* See Introduction, § 18.

, leader of

s step-son,

:ement of

is. Galba.

.oman war

: destruc-
d death of
Agricola :

est extent.

: emperors,
liogabalus,
ipire at its
A very few
Alexander
of Zenobia

A.D.

300. Tenth and last persecution of the Christians, under Diocletian and Galerius. Partition of the
empire amongst joint rulers, under the titles of
"Augusti" and "Cæsares." Abdication of
Diocletian and Maximian (305).

333. Constantine I. (the Great), son of Constantius
Chlorus, overcomes his five rivals, and obtains
the sole mastery of the Roman empire. He
makes Christianity the religion of the state, and
removes the seat of government from Rome to
Byzantium, thenceforth called Constantinople.
Council of Nice (the first œcumenic council) :
condemnation of the Arians (325).

375. Commencement of the great migrations : the im-
pulse given by the Huns (Mongols) in the East.
The Goths cross the Danube, and gain a victory
over the Emperor Valens at Adrianople. Translation of the Bible into Gothic by Ulfilas, bishop
of the Goths.

395. Theodosius the Great divides the empire betwixt
his two sons, Arcadius and Honorius, giving
the Eastern (Thrace, Asia Minor, Syria, and
Egypt) to the former, and the Western (Italy,
Africa, Gaul, Spain, and Britain) to the latter.
Ambrose, bishop of Milan : his two celebrated
scholars, Augustine and Chrysostom.

410. Rome plundered by Alaric the Visigoth. Carthage
conquered (429); and Rome plundered a second
time (455), by Genserich the Vandal. Vandal
kingdom in Africa. Suevi and Visigoths in

A.D.

410. Spain. Visigoths, Burgundians and Franks in Gaul. Saxons, Thuringians, and Alemanni in Germany.

450. Arrival in Britain of the Anglo-Saxons under Hengist and Horsa. Defeat of Attila, the Hun, at Chalons, by the Romans and the Visigoths (451).

476. Romulus Augustulus, the last Roman emperor of the West, dethroned by Odoacer, leader of the Heruli, who becomes king of Italy. End of the Western Roman Empire, and of the great migrations.

Franks in
lemanni in

ons under
, the Hun,
   Visigoths

mperor of
der of the
End of the
eat migra-

MEDIEVAL HISTORY.

A.D. 476—1517.

Medieval history embraces a period of more than a
thousand years, extending from the conquest of Teuto-
nism over Romanism in the Western Empire, to the
liberation of Northern Europe from the spiritual usur-
pation of Rome, by the Reformation. It exhibits the
northern nations becoming permeated by the civilisa-
tion of Greece and Rome ; their separation into Romanic and pure Teutonic nationalities ; but at the same
time the comprehension of both under the feudal and
hierarchical systems,—the former represented and head-
ed by the new German "Empire of the West," and the
latter by the waxing power of the papacy. Struggle betwixt these two powers. Central Europe becomes the
main theatre of events.

FIRST PERIOD.

A.D. 496—814.

FROM ODOACER TO THE DEATH OF CHARLEMAGNE.

End of the great migrations.—Rise of Mohammedan-
ism in the East.—Foundation of the new Empire of the
West.
A.D.
500. Theodoric the Great founds an Ostro-Gothic
empire in Italy. Chlodovic, first a heathen,

A.D.

500. afterwards a convert to Christianity, founds a
Frankish empire in Gaul.

555. Justinian I. Emperor of the East. Codification
of the Roman law. By his generals, Belisarius
and Narses, he overthrows the Vandal kingdom
of Africa, and the Ostrogothic of Italy.

568. Upper Italy conquered by the Longobardi (Lombards), under Alboin. Pavia their capital. Ra-
venna and the Exarchate remain Greek.

600. Gregory I. (the Great) Pope. Conversion of the
Anglo-Saxons.

622. Mohammed, "the prophet," founder of Islamism:
converts and conquers Arabia: his doctrine con-
tained in the Koran. His successors, the chalifs,
great conquerors in Syria, Persia, Egypt, and
Northern Africa.

711. Conquest of Spain by the Arabians: battle of
Xeres, and death of Roderick the last of the Goths.

732. Charles Martel, the Frankish Mayor of the Palace,
gains at Tours a great victory over the Arabs,
which arrests their progress in Europe.

752. Pepin the Short, son of Charles Martel, deposes
the last of the Merovingian "sluggard kings,"
the successors of Chlovis, and assumes the title
of King of the Franks. Diffusion of Christianity
in Germany by the labours of the English monk
Winifred or Bonifacius.

768 Charlemagne, son of Pepin, chief founder of the
to great Frankish power. His empire extends
814. from the Eider to the Tiber, and from the Ebro

founds a

dification
Belisarius
kingdom

rdi (Lom-
ital. Ra-
k.
ion of the

slamism:
trine con-
e chalifs,
gypt, and

battle of
he Goths.
e Palace,.
ne Arabs,

, deposes
l kings,"
the title
ristianity
ish monk

er of the
extends
the Ebro

A.D.
814. to the Raab. His wars against the Lombards, Arabians, and Saxons. His care of the church, of schools, and of science.
800. Charlemagne crowned Roman Emperor, on Christmas-day, by Pope Leo III.(+814.) His contemporary the caliph Haroun-al-Raschid.

SECOND PERIOD.

A.D. 814—1096.

FROM THE DEATH OF CHARLEMAGNE TO THE FIRST CRUSADE.

Struggles betwixt Christianity and Islamism in the East, and betwixt the ecclesiastical and secular powers in the West.—Commencement of the political preponderance of Germany in Europe.

A D.
814. Louis the Gentle (le Débonnaire) succeeds his father, Charlemagne: divides the empire with his sons: their rebellion and wars with each other.
843. Treaty of Verdun: division of the great Frankish empire. Lothaire receives the imperial title, Italy, and "Lorraine"; Charles (the Bald), France; and Louis (the German), the kingdom of Germany. The Grand Dukedom of Muscovy founded by Ruric.
871. Alfred the Great begins to reign in England (+901.)

A.D.

895. Rollo or Rolf sails up the Seine with a party of
Norman pirates ; establishes himself in Neustria,
thenceforth called Normandy; and embraces
Christianity.

919. Henry the Fowler, Emperor : he annexes Lorraine
to the empi.e, and extends his dominion to the
Oder.

987. Extinction of the Carlovingian kings by the death
of Louis V.   Accession of Hugh Capet, Duke of
Francia (whose dynasty reigns till 1328).

1000, Christianity now diffused over all Europe.

1017. Canute the Great begins his reign.   England
under Danish rule until 1041.

1024. Conrad II., first of the Franconian or Sali ; emperors of Germany.

1039. Henry III. Emperor. Imperial power at its height.
He tries to reform ecclesiastical abuses.   Expulsion of the Danes from England (1041).

1066. Norman invasion of England. Battle of Hastings.

1073 Gregory VI. (Hildebrand) elected Pope.   His
to   disputes with the Emperor Henry IV., whom he
1085. excommunicates.   He frees the church from subjection to temporal authority, and promulgates
the law of clerical celibacy.   At last Henry succeeds in deposing him (1084); and he dies in
exile the following year,

THIRD PERIOD.

A.D. 1096—1270.

FROM THE FIRST TO THE SEVENTH CRUSADE.

This is the most brilliant and distinctive period in the history of the Middle Ages. In the crusades we have the coalition of the two great elements of medieval life,—monachism and chivalry. In the contests of the House of Hohenstaufen with the Popes, we are presented with the second act of the great struggle between the civil and ecclesiastical powers, terminating in the victory of the latter.—Germany's political preponderance ends with the decay of the empire; the spiritual supremacy of Rome is established throughout Europe; but arts, science, and commerce flourish amongst the western nations.—The whole of Romanic and Teutonic Europe, together with a portion of Western Asia, is embraced in the theatre of events.

A.D.

1096 First crusade of the chivalry of Europe against
to the Saracens in Syria and Palestine, incited by
1099. the preaching of Peter of Amiens, ordered by
Pope Urban II., and led by Godfrey of Bouillon,
Duke of Lorraine. Some of the other leaders
were Godfrey's brother Baldwin, afterwards king
of Jerusalem, Raymond of Toulouse, Robert,
Duke of Normandy, son of William the Conqueror, and Tancred, the impersonation of chivalry. After the conquest of Jerusalem, foun-

ith a party of
lf in Neustria,
nd embraces

exes Lorraine
ainion to the

s by the death
apet, Duke of
1328).
Europe.
n. England

or Sali, em-

at its height.
ses. Expul-
)41).
of Hastings.
Pope. His
V., whom he
ch from sub-
promulgates
t Henry suc-
d he dies in

A.D.

1099. dation of the orders of Knights Templars and Knights of St. John.

1138. Conrad III. (+1152) commences the series of German emperors of the Suabian house of Hohenstaufen : undertakes, along with Louis VII. of France, the second crusade (1147–1149). Bernard of Clairvaux. Abelard, the scholastic philosopher. Arnold of Brescia.

1152. Frederick I.(Barbarossa),the Great Hohenstaufen. His power and influence in Italy and Germany. His wars with the Lombard cities, and his contests with the Pope. Vanquished in the great battle of Legnano (1177), he is forced to an accommodation with the Pope, and to allow the Lombard cities the right of self-government.

1155. Henry II. of England, the first Plantagenet. Thomas à Becket.

1190. Third crusade, under Frederick Barbarossa, Philip Augustus of France, and Richard I. of England. Death of Barbarossa : truce with Saladin : death of the latter (1193), during Richard's captivity in Germany.

1203. Fourth crusade. Revolution at Constantinople.

1215. Magna Charta granted by John Lackland. Innocent III., the most powerful of all the Popes in the Middle Ages. The Inquisition in the hands of the Dominicans, who, with the Franciscans, are called the begging orders. St. Francis of Assisi. Albigenses and Waldenses.

A.D.

1217 Fifth and sixth crusades,—the latter under the
to   Hohenstaufen Emperor Frederick II.   Genghis
1228. Khan establishes a great Mongolian empire, ex-
      tending from China to Russia.

1241. Progress of the Mongols in Europe arrested by
      the battle of Wahlstatt, near Liegnitz.   Com-
      mencement of the Hanseatic League.   Trouba-
      dours : Minnesingers.

1250. Frederick II. dies, excommunicated by the Pope,
      deposed by the Council of Lyons, but fighting
      to the last.   His death finishes the two hundred
      years' conflict betwixt the Church and the Em-
      pire : the former comes out victorious, and the
      latter falls into decay.   Commencement of the
      great interregnum in Germany, which lasts till
      the election of Rodolph of Hapsburg.

1265. First regular House of Commons assembled in
      England by Simon de Montford.

1270. Seventh and last crusade under Louis IX. of
      France (St. Louis).

FOURTH PERIOD.

A.D. 1270—1517.

FROM THE TERMINATION OF THE CRUSADES TO THE COM-
MENCEMENT OF THE REFORMATION.

This period exhibits the restoration of the German Empire in a more limited form than during the preceding
period.—A gradual decay of the papal power, ending

[left margin fragments]

Templars and

s the series of
a house of Ho-
rith Louis VII.
e (1147–1149).
, the scholastic

Hohenstaufen.
and Germany.
s, and his con-
d in the great
orced to an ac-
d to allow the
overnment.
; Plantagenet.

barossa, Philip
I. of England.
Saladin : death
ard's captivity

Jonstantinople.
Lackland. In-
f all the Popes
uisition in the
with the Fran-
orders.   St.
d Waldenses.

in its total subversion in the northern countries of Europe.—Decline of feudalism, of the hierarchy, and of monachism.—The German princes become politically and ecclesiastically independent of the empire; whereby the latter, losing its political unity, loses also its position as foremost of the European powers.—Flourishing condition of the cities in regard to commerce, industry, and the arts, until, by the discovery of America, trade is drawn away to the maritime towns.—A new direction given to the human mind by the invention of printing and the restoration of classical learning.

A.D.

1273. Rodolph of Hapsburg chosen emperor. The empire begins to recover from the confusion of the interregnum. Rodolph defeats Ottocar, King of Bohemia, deprives him of the Duchy of Austria, and confers it on his own son Albert.

1282. Insurrection of Sicilians against Charles of Anjou: "Sicilian vespers."

1305. Bertrand, Archbishop of Bordeaux, being elected Pope (Clement V.), removes the papal see to Avignon, where it remains for a period of seventy years, called by the Italians the "Seventy years' captivity."

1308. The Helvetic cantons of Schwytz, Uri, and Unterwalden revolt against Albert of Hapsburg, and lay the foundation of the Swiss confederation: Stauffacher, Fürst, Melchthal, William Tell. Their wars with Austria: battle of Morgarten (1315).

1356. The Golden Bull, settling the constitution of the

intries of Eu-
irchy, and of
ne politically
pire; whereby
also its posi-
—Flourishing
rce, industry,
merica, trade
A new direc-
invention of
earning.

iperor. The
confusion of
Ottocar, King
ichy of Aus-
Albert.
Charles of

being elected
papal see to
od of seventy
eventy years'

Uri, and Un-
apsburg, and
nfederation :
'illiam Tell.
if Morgarten

itution of the

A.D.

1356. German Empire, and vesting the choice of the
emperor in the seven "Electors," issued by
Charles IV. Battle of Poictiers : Edward, the
Black Prince.

1386. Battle of Sempach gained by the Swiss.

1397. Union of Calmar : Succession of the three Scandinavian kingdoms, then ruled by Margaret,
"the Semiramis of the North," settled on her
grand-nephew.

1414. Council of Constance convened at the instance
of the Emperor Sigismund. Burning of John
Huss and of Jerome of Prague. Battle of Agin-
court (1415), where Henry V. defeats a French
army eight times as numerous as his own.

1429. Joan of Arc raises the siege of Orleans. Charles
VII. crowned at Rheims. Council of Basle
(1431).

1453. Constantinople taken by the Turks : end of the
Eastern Roman Empire.

1455. The first printed book—a copy of the Vulgate—
issues from the press: printed by John Gutenberg,
John Fust, and Peter Schæffer. Wars of the
Roses in England (1454–1484).

1477. Austria acquires Burgundy by the marriage of
Maximilian with Mary, daughter and heiress of
Charles the Rash. Louis XI. of France.

1492. Discovery of America by Columbus. Expulsion
of the Moors from Spain. Ferdinand and Isa-
bella.

1498. Discovery of sea-route to India, round the Cape
of Good Hope, by Vasco de Gama, a Portuguese.

MODERN HISTORY.

A.D. 1517—1859.

Modern History commences with the breaking up of the feudal system and of the Romish hierarchy.—The religious life which is the immediate product of the Reformation, by degrees becomes colder, and interests of a more secular sort acquire the ascendancy in politics.—The range of European influences is extended beyond the limits of the old world by the establishment of colonies in the new.—Meantime culture is progressive in Europe, especially towards the North and the East.—The Slavonic race begins to make a figure in history.—Germany continues to decline as a political power, and is split up into numerous sections with separate, and frequently opposing, interests.—The monarchical form of government attains its highest development, to be at length overthrown by revolution and afterwards partially restored, whilst the commons are continuously growing in education, intelligence, wealth, and political importance. Science, and the knowledge of classical antiquity, are progressive.

FIRST PERIOD.

A.D. 1417—1648.

FROM THE REFORMATION TO THE PEACE OF WESTPHALIA.

Ecclesiastical separation of Northern (Protestant) from Southern (Catholic) Europe.—Embittered contest,

reaking up of
rarchy.—The
oduct of the
and interests
lancy in poli-
s is extended
the establish-
culture is pro-
he North and
ake a figure in
as a political
sections with
terests.— The
s its highest
by revolution
the commons
, intelligence,
ence, and the
gressive.

WESTPHALIA.

(Protestant)
tered contest,

by sword and pen, of the two ecclesiastical and political parties.—Political preponderance of the House of
Spain and Austria.—Religio-political character of the
period.—Total dissolution of German unity into a
number of petty states, composing a nominal Empire,
and decline of its material prosperity in consequence of
the altered direction given to commerce by the discovery of America, and of the route to India by the Cape of
Good Hope.—England, in conflict with Spain, lays the
foundation of her greatness by directing her energies
to the attainment of maritime superiority; and, after becoming united with Scotland into one kingdom, passes
through a period of domestic convulsion which results
in a temporary overthrow of the monarchy.

A. D.
1517. Commencement of the Reformation; in Germany
by Luther's 95 theses against indulgences; in
Switzerland by Zwingle and Calvin. Leo. X.
Henry VIII.
1521. Luther before the diet of Worms, presided over
by Charles V. His imprisonment in the Wartburg.
Translation of the Bible. Raphael.
1525. Battle of Pavia: Francis I. of France defeated
and taken prisoner: " All is lost but honor."
1530. Diet of Augsburg. Melanchthon. Confession of
Augsburg.
1531. Henry VIII. proclaimed supreme head of the
English Church : suppresses the monasteries
(1535).
1540. The order of the Jesuits founded by Ignatius
Loyola. It becomes the most important engine
for opposing the Reformation.

E

A.D.

1543. Copernicus of Thorn propounds a new system of
astronomy.

1555. Religious peace of Augsburg: liberty of doctrine
and worship guaranteed to German Protestants.
Abdication of Charles V. severs connection between Spain and Germany. First *Index Expur-
gatorius* published by Paul IV.

1563. Church of England established on its present
basis: thirty-nine articles. Dissolution of the
Council of Trent (assembled originally 1545),
and formal ratification of its decrees by Pope
Pius IV.

1566. Revolt of the Netherlands: Cruelties of the Duke
of Alva: Execution of Counts Egmont and Horn
with 18,000 others: Foundation of the Republic
of the Seven united Provinces, with William
Prince of Orange (William the Silent) as Statt-
holder (1579).

1572. Paris massacre of St. Bartholomew.

1582. Pope Gregory XIII. reforms the calendar.

1587. Execution of Mary, Queen of Scots. Equip-
ment of the Invincible Armada: its destruction
(1588).

1589. Accession of the House of Bourbon in France:
Henry IV. (Henry of Navarre): Sully.

1598. Edict of Nantes (by which liberty of conscience
guaranteed to the French Protestants) issued by
Henry IV. Death of Philip II.

1603. Death of Queen Elizabeth: Accession of James
I. unites the English and Scotch crowns. Shakes-
peare. Cervantes.

new system of

rty of doctrine
m Protestants.
connection be-
Index Expur-

on its present
olution of the
ginally 1545),
crees by Pope

ies of the Duke
mont and Horn
f the Republic
with William
lent) as Statt-

r.
alendar.
cots. Equip-
its destruction

on in France:
ully.
of conscience
nts) issued by

sion of James
owns. Shakes-

A.D.

1613. Accession of the House of Romanoff in Russia.

1618. Outbreak of the thirty years' war. Elector Palatine defeated at Prague (1620), and placed under the ban of the empire. Tilly. Wallenstein.

1630. Gustavus Adolphus, King of Sweden, takes the field in Germany as champion of the Protestant cause. Battle of Leipzig: the Swedes victorious: Tilly slain.

1632. Battle of Lützen: the Swedes victorious: Gustavus slain. Murder of Wallenstein (1634). France governed by Richelieu. Charles I. attempts to levy taxes in England without the consent of Parliament. Ship-money. Hampden.

1640. Assembling of the Long Parliament. Trial and execution of Laud and Strafford. Episcopacy abolished. Charles takes the field: battle of Edge-hill. Oliver Cromwell.

1647. Charles a prisoner. Dissolution of the Long Parliament by Cromwell (1648): King's trial and execution (1649).

1648. Peace of Westphalia (Treaties of Münster and Osnabrück). Religious freedom guaranteed to the Lutheran and Reformed Churches in all the German States. The Emperor's power restrained within still narrower limits, and Switzerland and the United Provinces formally recognised as independent republics.

SECOND PERIOD.

A. D. 1648—1789.

FROM THE PEACE OF WESTPHALIA TO THE OUTBREAK OF
THE FRENCH REVOLUTION.

The interests of commerce become predominant in
politics.—The monarchical form of government attains
its culminating point, especially in France, now become
the first power of the European continent.—Rise of
Prussia and Russia in conflict with Sweden.—Austria
engaged in hostilities with the Turks.—England and
Holland become great naval powers.—The religious
excitement of the first period gives place to laxity or
"liberalism."—The growth of revolutionary ideas in the
British American Colonies forms a prelude to the overthrow of absolute monarchy in Europe.—Under Frederick the Great, Prussia becomes one of the great
powers.

A. D.

1649. Cromwell protector in England. France gov-
      erned by Cardinal Mazarin, Richelieu's successor.
      Wars of the Fronde.

1660. Restoration of the Stuarts. On the death of
      Mazarin, the actual exercise of government is
      assumed by Louis XIV., the "grand monarque,"
      with notions of absolute power: "L'état c'est
      moi." Colbert, Turenne, Condé.

1683. Vienna, besieged by the Turks, is relieved by
      John Sobieski, king of Poland.

TBREAK OF

lominant in
nent attains
now become
it.—Rise of
n.—Austria
ngland and
he religious
to laxity or
ideas in the
to the over-
-Under Fre-
f the great

France gov-
r's successor.

he death of
vernment is
monarque,"
'L'état c'est

relieved by

A.D.

1685. Revocation of the Edict of Nantes by Louis XIV. Consequent emigration of French Protestants, who carry their arts and industry to England, Holland, and Germany.

1688. English revolution. Abdication of James II. William and Mary joint sovereigns.

1697. Peter the Great sets out on his travels. European tactics introduced into the Russian army by General Gordon.

1700. Commencement of the great northern war between Charles XII. and Peter the Great. Battle of Narva.

1701. The Elector of Brandenburg assumes the title of King of Prussia. Commencement of the war of the Spanish succession. Prince Eugene. Duke of Marlborough. Battles of Blenheim (1704), Ramillies (1706), Malplaquet (1709).

1713. Treaty of Utrecht: Gibraltar, Minorca, Nova Scotia, Newfoundland, Hudson's Bay Territory, and St. Christopher's ceded to Great Britain.

1714. Queen Anne dies: accession of the House of Brunswick in the person of George, Elector of Hanover.

1715. Death of Louis XIV.: succeeded by his great-grandson, Louis XV.: Philip, Duke of Orleans regent. Rising in Scotland in favor of the exiled Stuarts.

1721. Peter the Great, hitherto Czar of Muscovy, assumes the title of "Emperor and Autocrat of all the Russias." +1725.

A.D.

1733. War of the Polish succession. Establishment of the Bourbons at Naples (1735). The Duke of Lorraine receives the Grand Duchy of Tuscany.

1740. Accession of Frederick II. (the Great) to the Prussian throne. War of the Austrian succession: Hungarian diet: "Moriamur pro rege nostro Maria Theresia."

1745. Rebellion in Scotland under Charles Edward. The Grand Duke of Tuscany chosen Emperor of Germany (Francis I.). Battle of Culloden (1746). Voltaire. Rousseau.

1752. New style adopted in Great Britain.

1756. Commencement of the seven years' war. William Pitt, afterwards Earl of Chatham, minister in England. The French dispossessed in America (Wolfe), and in India (Clive).

1762. Catharine II. becomes Empress of Russia. Termination of the seven years' war. Britain retains Canada, Cape Breton, the islands and coasts of the St. Lawrence, and the territories on the left bank of the Mississippi.

1773. First partition of Poland. Suppression of the Jesuits by Clement XIV. (Ganganelli). Treaty of Kainardji between Russia and Turkey (1774), from which dates the great influence gained by the former state in the affairs of the latter. Warren Hastings Governor-general of India.

1775. Commencement of the War of Independence in the British North American Colonies: George Washington : Benjamin Franklin.

ablishment of
The Duke of
of Tuscany.
treat) to the
trian succes-
ur pro rege

rles Edward.
a Emperor of
loden (1746).

l.
war. William
t, minister in
d in America

of Russia.
war. Britain
islands and
he territories

ession of the
lli). Treaty
urkey (1774),
ce gained by
f the latter.
of India.
pendence in
ies : George

A.D.

1783. Conclusion of the American war. Great Britain acknowledges the independence of the colonies now forming the federal republic of the United States.

THIRD PERIOD.

A. D. 1789—1859.

FROM THE OUTBREAK OF THE FRENCH REVOLUTION TO THE PRESENT DAY.

The first twenty-five years of this period (1785-1815) is a period of revolutionary ferment.—In France the monarchical form of government is overthrown, to make way for the anarchy and despotism of the revolution: the name of freedom is frightfully abused: great part of Europe is laid prostrate at the feet of a military adventurer, but at last rises and dethrones him.—Subsequently, we perceive the formation of constitutional governments on the European continent; the establishment of a system of political equilibrium under the five great powers, viz. England, Russia, France, Austria, and Prussia; a state of continued warfare betwixt conservatism and radicalism in church and state, along with agricultural, industrial, and commercial prosperity, especially in England.— The United States of America become a great republican power, and a constantly increasing stream of emigration sets in from the Old World to the New.

A. D.

1789. Convocation of the "States-General," which afterwards resolve themselves into the "Constituent Assembly." Necker, Mirabeau, Duke of Orleans. Bastille taken. National Guard; Lafayette. Abolition of nobility and titles Emigration.

1790. France divided into departments. Crown lands and church property confiscated. Assignats.

1791. King's flight to Varennes. Legislative Assembly. Constitutional royalists; Girondists, Mountain, Jacobins, Guillotine.

1792. Storming of Tuilleries: King and Royal family sent as prisoners to the Temple: September massacres. National convention. Commencement of the revolutionary war. Battles of Valmy and Jemappes.

1793. King and Queen beheaded. Overthrow of the Girondists. Reign of terror. Revolutionary tribunal. Robespierre, Marat, Danton. Worship of reason. Revolutionary calendar. War declared by France against England, which forms alliances with and subsidizes the other continental powers. Second partition of Poland.

1794. Overthrow of Robespierre and the terrorists. Rising of the Poles under Kosciuzeo.

1795. The Directory. Conquest of Holland by Pichegru. Batavian Republic. Third partition of Poland.

1796. Napoleon Bonaparte's Italian campaign: battle of Lodi: siege of Mantua; battles of Arcola and Rivoli.

A.D.

1797. Surrender of Mantua. Peace of Campo Formio. Dissolution of the Venetian, and formation of the Cisalpine and Ligurian republics.

1798. Bonaparte's expedition to Egypt. Battle of Aboukir : destruction of the French fleet by Nelson. Pope carried prisoner to France. Italy and Switzerland revolutionized.

1799. Second coalition against France. Austria and Russia take the field. Suwarrow's campaign in Italy. Bonaparte's return from Egypt. Overthrow of the Directory. Bonaparte first consul.

1800. Battle of Marengo, won by Bonaparte, and of Hohenlinden, won by Moreau.

1801. Peace of Luneville : whole left bank of Rhine ceded to France. Secularizations in Germany. Paul of Russia murdered, and succeeded by Alexander I.

1802. Peace of Amiens betwixt France and England.

1803. Renewal of hostilities betwixt France and England. Occupation of Hanover. Abolition of ecclesiastical sovereignties in Germany.

1804. Napoleon proclaimed Emperor of the French. Crowned the same year.

1805. Napoleon crowned King of Italy. Third coalition against France, composed of Great Britain, Austria, Russia, and Sweden. Ulm surrendered by General Mack. Battle of Austerlitz compels Austria to the peace of Pressburg. Battle of Trafalgar and death of Nelson.

A.D.

1806. Confederation of the Rhine. Dissolution of the
German Empire. Francis renounces the title of
Emperor of Germany, and assumes the new title
of Emperor of Austria. Joseph Bonaparte is
made King of Naples, and Louis Bonaparte King
of Holland. Battle of Jena.

1807. Battles of Eylau and Friedland. Peace of Tilsit.
Jerome Bonaparte King of Westphalia. Continental system. Bombardment of Copenhagen.
Portugal occupied by the French: Portuguese
court removes to Brazil.

1808. Napoleon's interview with Charles IV. and Ferdinand VII. at Bayonne. Joseph Bonaparte
king of Spain. Murat king of Naples. Outbreak
of the Spanish war: Wellesley (Wellington)
victorious in Portugal. Congress of Erfurt.

1809. Austria renews hostilities against France. Battles of Eckmühl, Aspern, and Wagram. Rising
of the Tyrolese under Hofer. Peace of Schön-
brunn. Kingdom of Illyria ceded to France.
Walcheren expedition. Battle of Talavera.
Revolution in Sweden: Gustavus IV. deposed,
and his uncle, Charles XII., chosen in his stead.

1810. Napoleon at the height of his power; his divorce
from Josephine, and marriage with Maria Louisa,
Arch-Duchess of Austria. The States of the
Church, Holland, and Northern Germany united
to the French empire. Bernadotte Crown Prince
of Sweden. Wellington in Spain: lines of
Torres Vedras: siege of Cadiz: Revolt of Spanish
South American Colonies.

lution of the
s the title of
the new title
onaparte is
aparte King

ace of Tilsit.
lia. Conti-
openhagen.
Portuguese

V. and Fer-
Bonaparte
Outbreak
Vellington)
Erfurt.
nce. Bat-
m. Rising
of Schön-
to France.
Talavera.
. deposed,
his stead.
his divorce
ria Louisa,
es of the
any united
wn Prince
lines of
f Spanish

A.D.

1812. Napoleon's expedition against Russia : the Grand Army : Smolensk : battle of Borodino : Kutusow : burning of Moscow : Rostopschin : Napoleon forced to retreat : battle of Malo-Jaroslawitz : passage of the Beresina : annihilation of the Grand Army. Wellington in Spain : battle of Salamanca.

1813. Germany rises against Napoleon : liberation war : Austria and Sweden join the coalition : Schwarzenberg : Blücher. Battles of Dresden, Culm, and Leipzig. Dissolution of the Rhenish confederacy. Peace of Kiel : Norway ceded by Denmark to Sweden. Wellington gains the battle of Vittoria.

1814. The allies enter France : battles of La Rothière, Montmirail, Laon, and Arcis-sur-Aube : Paris taken : battle of Toulouse : Napoleon abdicates at Fontainebleau, and retires to Elba. First peace of Paris : France confined to her boundaries in 1792 : restoration of the Bourbons : Louis XVIII. : the Charter. Congress of Vienna : Austria receives an accession of territory in Italy.

1815. Kingdom of the Netherlands : Abolition of the slave trade. Constitution of Germany as a Confederation of thirty-eight states. Napoleon escapes from Elba and lands in France : the hundred days : battles of Quatre Bras and Waterloo : Wellington, Blücher. Paris occupied by the allies. Second restoration of Louis XVIII.

A.D.

1815. Napoleon sent prisoner to St. Helena, where he dies in 1821.   Holy alliance betwixt Austria, Russia, and Prussia.

1818. Congress of Aix-la-Chapelle : France joins the holy alliance.

1820. Revolutions in Spain, Portugal, Naples, and Piedmont.   Congress of Troppau.   Death of George III.

1821. Great insurrections in Moldavia, Wallachia, and the Morea.   Revolution in Brazil : Don Pedro Emperor.

1822. Congress of Verona.   Philhellenes : Lord Byron in Greece.   Canning appointed foreign secretary on the death of Lord Londonderry.

1823. French intervention in Spain : absolutism restored there and in Portugal.   Civil war in Greece.

1824. Death of Louis XVIII., and accession of his brother Charles X.

1825. Alexander, Emperor of Russia, dies at Taganrog. Commotions at St. Petersburg : accession of Nicholas.

1827. Intervention of England, France, and Russia in the affairs of Greece : battle of Navarino. Premiership and death of Canning.

1828. Duke of Wellington premier.   Russian invasion of Turkey.   Capo d'Istria, President of Greece : a French army in the Morea.   Don Miguel usurps the throne of Portugal.

1829. The Russian Field-Marshal Diebitsch crosses the Balkan.   Treaty of Adrianople.   Indepen-

lena, where he
twixt Austria,

ince joins the

Naples, and
u. Death of

allachia, and
: Don Pedro

Lord Byron
ign secretary

atism restor-
r in Greece.
sion of his

t Taganrog.
ccession of

d Russia in
- Navarino.

n invasion
of Greece :
ue usurps

h crosses
Indepen-

A.D.
1829. dence of Greece recognized by Turkey. Catholic emancipation in England.

1830. Accession of William IV. Algiers taken by the French. July revolution at Paris : abdication of Charles X. : Duke of Orleans called to the throne, by the title of Louis Philippe King of the French. Belgian and Polish revolutions.

1831. The cholera appears in Europe. Polish insurrection suppressed, and the kingdom of Poland incorporated with the Russian empire. London conferences : Leopold of Saxe-Coburg chosen King of Belgium.

1832. Civil war in Portugal betwixt Pedro and Miguel. The French occupy Ancona, and lay siege to Antwerp. Parliamentary reform in England.

1833. Meeting of the first reformed Parliament. Abolition of slavery in the British colonies, with a compensation of £20,000,000 to the slave-owners.

1834. Don Miguel expelled from Portugal. Civil war in Spain. Formation of the German Zollverein.

1837. Accession of Queen Victoria. Insurrection in Canada.

1839. Treaty of peace betwixt Holland and Belgium. End of the civil war in Spain.

1840. Intervention of England and Austria in the Egyptian question. Thiers minister in France : apprehensions of a general war : removed by the overthrow of Thiers : Guizot minister.

1841. Resignation of Melbourne ministry. Peel becomes premier. Death of Lord Sydenham in Canada. Fortification of Paris.

A.D.

1842. Afghan and Chinese wars: cession of Hong
Kong to England: opening of Chinese ports.
Rising against the English at Cabul: murder of
Burnes and McNaughten: massacre at the Cabul
Pass. Ashburton Treaty with the United States.
Great fire at Hamburg.

1843. Activity of the Anti-Corn Law League. John
Bright returned for Durham. Queen Victoria
and Prince Albert visit the King of the French
and the King of the Belgians. Repeal meetings
in Ireland stopped by royal proclamation, and
Mr. O'Connell and other repealers arrested and
tried for conspiracy and sedition.

1844. French hostilities with Morocco: Mogadore
bombarded: King of the French visits Queen
Victoria at Windsor. Railway mania in England.

1845. Continued activity of the Anti-Corn Law League.
Great bazaar at London, where the receipts
amount to £25,000. Railway mania in England
attains its height: scrip issued to the nominal
amount of several hundred millions sterling.
Annexation of Texas to the United States. Steam
established between Liverpool and New York.

1846. The Spanish double marriages. Coolness betwixt the courts of St. James and the Tuilleries.
Abolition of the Corn Laws, followed by resignation of the Peel ministry. Austria, in violation
of the treaties of Vienna, seizes on Cracow, and
incorporates it with her own dominions. Louis

A.D.

1846. Napoleon escapes from the Castle of Ham, in Normandy. Gregory XVI. dies, and is succeeded by Cardinal Mastai Ferretti, who takes the title of Pius IX.

1847. Pope Pius introduces some reforms into the Papal States: excitement in the rest of Italy. Civil war in Switzerland: Sonderbund suppressed. Abd el Kader taken prisoner. The Duchy of Lucca reverts to Tuscany.

1848. Upper California and New Mexico ceded to the United States. February revolution in Paris: flight of Louis Philippe; France a Republic: Cavaignac: Louis Napoleon President. Revolutions at Vienna and Berlin. Schleswig-Holstein insurrection.

1849 Revolutions in Rome and Tuscany: Mazzini; French invasion and occupation of Rome. Revolutionary movements in Germany and Hungary. Kossuth. Revolution in Baden suppressed by Prussia; in Hungary by Russia.

1850. Battle of Idstedt and suppression of the Schleswig Holstein insurrection. Peace between Denmark and Prussia.

1851. Great industrial exhibition in London. French coup d'état: National Assembly broken up, and Napoleon declared President of the Republic for ten years. Discovery of gold fields in Australia.

1852. The Earl of Derby forms a protectionist ministry, dissolves parliament, but is soon forced to

(left margin fragments) on of Hong / inese ports. / : murder of / at the Cabul / nited States. / gue. John / en Victoria / the French / al meetings / nation, and / rested and / Mogadore / Its Queen / in Eng- / League. / receipts / England / nominal / sterling. / Steam / York. / ness be- / illeries. / y resig- / iolation / ow, and / Louis

A.D.

1852. resign : Lord Aberdeen becomes Premier. Louis
      Napoleon proclaimed Emperor of the French, as
      Napoleon III.
1853. Russia invades the Danubian principalities;
      destroys Turkish fleet at Sinope.
1854. Great Britain and France declare war against
      Russia. Landing of the allies at Varna : they
      invade the Crimea. Battles of Alma, Balaklava,
      Inkermann. Siege of Sebastopol.
1855. Sardinia joins the allies : south side of Sebas-
      topol taken : Russia proposes peace. Treaty of
      peace signed at Paris the following year.